Sediment Transport in Irrigation Canals: A New Approach

Sediment Transport in Irrigation Canals: A New Approach

Editor

Shivaji Shende

scitus
academics

Sediment Transport in Irrigation Canals: A New Approach

Edited by **Shivaji Shende**

Printed in 2017

ISBN: 978-1-68117-009-1

Library of Congress Control Number: 2015931817

SCITUS Academics LLC,
616, Corporate Way, Suite 2, 4766,
Valley Cottage, NY 10989

www.scitusacademics.com

Contents

Preface

An irrigation canal is a waterway, often man-made or enhanced, built for the purpose of carrying water from a source such as a lake, river, or stream, to soil used for farming or landscaping.

The engineering concept of efficiency at different levels of a canal system is important for planning, designing and even operating a canal irrigation system.

Much more challenging than that of groundwater is the improvement of performance of canal systems, which will remain the dominant water transfer technique in developing countries in the future. The standards used for design of irrigation canals in many countries have not kept up with the development of new technologies. Most traditional delivery systems have no or little flexibility built into them. Their control strategies and control structures are inadequate for good water management practices. Water deliveries at farm level are not compatible with the key requirements of modern agriculture. Thoughts on how to assess the performance of canal projects are evolving.

Efficiency alone is not a sufficient indicator to define the performance of an irrigation system. Indicators of water productivity expressed in terms of production or value of production per unit of water are very important in increasingly water-scarce situation. A canal irrigation system may have high conveyance efficiency with a minimum of seepage and operational losses. However, if water delivery is too rigid or unreliable, there will be considerable waste further down at the farm level. A water productivity indicator provides a global indication of the effectiveness of water conservation measures and of the quality of service provided to the users, as well as the farm use of water and other inputs.

This book gives an understanding of the behaviour of sediments in irrigation canals. This book will also help to decide on the appropriate design of the system, the water delivery plans, to evaluate design alternatives and to achieve an adequate and reliable water supply to the farmers.

Editor

Chapter 1

Potential Use of Organic- and Hard-Rock Mine Wastes on Aided Phytostabilization of Large-Scale Mine Tailings under Semiarid Mediterranean Climatic Conditions: Short-Term Field Study

Claudia Santibañez,[1,2] Luz María de la Fuente,[2] Elena Bustamante,[2] Sergio Silva,[3] Pedro León-Lobos,[3], and Rosanna Ginocchio[2]

[1]Facultad de Ciencias Silvoagropecuarias, Universidad Mayor, Camino La Pirámide 5750, Huechuraba, Santiago, Chile

[2]Unidad de Fitotoxicidad y Fitorremediación, Centro de Investigación

Minera y Metalúrgica, Avenida Parque Antonio Rabat 6500, Vitacura 7660045, Santiago, Chile

[3]Instituto de Investigaciones Agropecuarias, CRI Intihuasi, Colina San Joaquín s/n, La Serena, Chile

ABSTRACT

The study evaluated the efficacy of organic- and hard-rock mine waste type materials on aided phytostabilization of Cu mine tailings under semiarid Mediterranean conditions in order to promote integrated waste management practices at local levels and to rehabilitate large-scale (from 300 to 3,000 ha) postoperative tailings storage facilities (TSFs). A field trial with 13 treatments was established on a TSF to test the efficacy of six waste-type locally available amendments (grape and olive residues, bio solids, goat manure, sediments from irrigation canals, and rubble from Cu-oxide lixiviation piles) during early phases of site rehabilitation. Results showed that, even though an interesting range of waste-type materials were tested, bio solids (100 t ha^{-1} dry weight, d. w.) and grape residues (200 t ha^{-1} d.w.), either alone or mixed, were the most suitable organic amendments when incorporated into tailings to a depth of 20 cm. Incorporation of both rubble from Cu-oxide lixiviation piles and goat manure into upper tailings also had effective results. All these treatments improved chemical and microbiological properties of tailings and lead to a significant increase in plant yield after three years from trial establishment. Longer-term evaluations are, however required to evaluate self-sustainability of created systems without further incorporation of amendments.

INTRODUCTION

Copper mining operations may adversely affect the environment due to deposition of large volumes of a number of hard-rock waste materials in nearby areas, such as sterile rocks, smelter slags, smelter dust, and tailings, among others. When sulfide copper ores are concentrated by flotation, approximately 80% of total wastes are tailings, which still contain a concentration of metals (i.e., Cu, Zn, Mo, Ni, Pb, Cd) and metalloids (i.e., As) that may pose environmental risks after inadequate deposition and management [1–6]. Currently, tailings are deposited in artificial dumps, or tailings storage facilities (TSFs), where fine solid particles (tailing sands) are separated from water by gravity [7]. Abandonment of postoperative TSFs under semiarid Mediterranean climatic conditions, such as in north-central Chile, led to complete water evaporation from upper tailings [7]. Without proper closure management, this fine, homogenous, and noncohesive material is left exposed to physical and chemical environmental forces [1, 8], causing erosion by wind. Deposition of metal/metalloid-rich tailings into nearby soils and surface waters may pose risks to human health, agricultural activities, and wildlife [3, 9, and 10]. Depending on tailings mineralogy (i.e., content of indigenous iron-/sulfur-oxidizing bacteria [11, 12]), and geographic location, acid mine drainage and lixiviation of metals may also occur with some secondary environmental impacts on surface and ground waters [7].

After a long period since abandonment, postoperative TSFs are generally devoid of vegetation or covered with a rather scarce vegetation cover [13–15]. Spontaneous plant colonization of TSFs is a very slow process as mine tailings are usually characterized by bad drainage, compactation, absence of organic matter and nitrogen, neutral to low pH [7, 14, 16], and limited soil-type microbiota to make energy and nutrients of litter available to plants [10]. Additionally, under semiarid climatic conditions, these problems are exacerbated by the builtup of salinity, as a high proportion of rainfall and the water contained in TSFs undergo evaporation rather than infiltration [7, 13]. Therefore, physical and chemical characteristics of mine tailings, among other factors, interact to almost completely suppress seed germination, plant growth, and microbial activity [10, 17].

Aided phytostabilization is recognized as a potentially cost-effective and ecologically sound approach to containment of metal-polluted soils and mine tailings [5, 15, 18–20]. Their primary objectives are to reduce the mobility, ecotoxicity, and dispersion of metals/metalloids through the environment [5, 15, 19, and 20]. This technique requires both incorporation of adequate inorganic and/or organic amendments to topsoils and revegetation with adequate plant species (i.e., metal tolerant plants, likely with an excluder phenotype) for immobilizing metals/metalloids in the rhizospheric zone and providing erosion control and wildlife habitat [2–5, 5–22]. Use of native plants is a focus of this technology because they often demonstrate tolerance to local environmental conditions and provide a foundation for natural ecological succession. However, success of aided phytostabilization on mine tailings also depends on the rapid improvement of most, if not all, limiting physical and chemical factors of the substrate for proper plant establishment and growth, including the establishment of proper soil microbial activity [16, 23–25]. Microbial activity is fundamental for the biochemical cycling of carbon, nitrogen, and phosphorus and the processes of organic matter turnover [26], and thus for achieving self-sustainable ecosystems [27].

Establishment of a vegetation cover helps to control erosion and provide organic matter in the substrate, increasing substrate aggregation and sequestration of pollutants [5, 15]. Most of these functions can be induced through proper application of amendments into mine tailings, such as organic and/or inorganic materials. In general, inorganic amendments improve either physical characteristics of tailings, such as compactation and drainage, or some limited chemical characteristics, such as pH and excessive soluble metal levels [25, 28–30]. Organic amendments improve physical characteristics of tailings, mitigate their metal toxicity, inoculate them with soil microorganisms, and incorporate required macronutrients and organic matter [5, 15, 25, 27, 28, 31–33].

Even though aided phytostabilization has been effectively used on metal-enriched soils and hard-rock mine wasted, such as TSFs, particularly in temperate areas (i.e., [10, 23, 24, 34, 35]), its use on postoperative TSFs of north-central Chile represent a new challenge. On one hand, postoperative TSFs being generated by large-scale copper mine operations in Chile have larger surface areas than the ones generated in other mining countries (range of 300 to 3,000 ha); therefore, local availability of very large amounts of proper amendments is by far a more limiting aspect for application of the technology than in other copper mining countries. This represents an opportunity for generating alternative disposal areas for other massive wastes that cannot be largely applied on croplands (i.e., agrowastes, biosolids, hard-rock mine wastes); however, their efficacy to improve physical, chemical, and biological properties of tailings have to be evaluated. On the other hand, experiences of aided phytostabilization on TSFs mainly gained in temperate climates (i.e., Canada) may not be directly applied to TSFs located in semiarid regions, such as north-central Chile. For example, organic covers (approx. 50 cm depth) made with biosolids and/or other organic wastes generated by cellulose-producing plants or agricultural activities have been effectively used on aided phytostabilization programs on TSFs in temperate areas (i.e., [10, 23, 24, 34, 35]), but they are inadequate under Mediterranean semiarid climate type conditions as salinization of organic covers strongly limit plant establishment and growth (i.e., [5, 25, 36, 37]). Therefore, alternative management options, such as incorporation into upper tailings, have to be evaluated.

The main objective of the present study was to assess the efficacy of a range of locally available organic and hard-rock mine waste-type materials on aided phytostabilization of postoperative copper TSFs under semiarid Mediterranean climate type conditions, with emphasis on chemical and biological parameters. The use of locally available wastes as tailing amendments, through promotion of integrated waste management practices at local level, was prioritized in order to achieve the large volumes that will be required by the large-scale copper mine operations in north-central Chile. Plant yield, metal uptake and translocation to aerial tissues, and the evolution of a number of microbiological and chemical parameters of the substrate were assessed.

MATERIALS AND METHODS

Study Site

The study was conducted at La Cocinera TSF (6.7 ha; 6.618.700 N-291.300 E) owned by the Ovalle copper mining plant of ENAMI. This postoperative and dried TSF is located in north-central Chile, Coquimbo Region, in an area under semiarid Mediterranean climate type conditions. Annual rainfall in the area averages 237 mm, concentrated during April to September (Autumn-Winter). The dry season extends from October to May. During this period the water balance is negative, resulting in a soil moisture deficit. Mean temperature of the warmest month (January) and the coldest month (July) are 23°C and 7°C, respectively [36].

Tailings of La Cocinera TSF are characterized by a clay loam texture, slightly alkaline pH (7.98), elevated electric conductivity (EC, 5.62 mS m^{-1}), very low organic matter (0.48%) and available N-P contents, low cation-exchange capacity (CEC, 6.28 meq 100 mg^{-1}), and very high concentrations of sulphate (2,912 mg L^{-1}), total copper (4,393 mg kg^{-1}, dry weight basis (d.w.)), total zinc (1,619 mg kg^{-1} d.w.), total iron (87,094 mg kg^{-1} d.w.), and total calcium (35,504 mg kg^{-1} d.w.; [15]). The existing literature [38, 39] describes the local soils as clay loam in texture, from colluvial and alluvial origin, neither saline (EC of 2.29 mS m^{-1}) nor sodic, with organic matter content around 1.5–2.8%, pH of 7.42, and CEC of 13.27 meq 100 mg^{-1}. Even though, some physical and chemical characteristics of study tailings are similar to local soils, such as texture and pH, most of the others need to be improved in order to support a native plant cover. Therefore, the main emphasis of the present study was to evaluate the efficacy of selected amendments to improve limiting chemical and biological parameters of tailings.

Experimental Design

A 2,400 m^2 field trial (60 m long × 40 m wide) was conducted at La Cocinera TSF to evaluate some biological and chemical endpoints of mine tailings amended with several local available organic and inorganic residues for a total of 12 treatment plots and a control (no amended tailings) plot (Table 1); each treatment was replicated three times. Selected amendments were discarded rubble from Cu-oxide lixiviation piles (R), available at the same mine operation; goat manure (M) from nearby cattle yards; sediments from the cleaning process of local irrigation canals (S); air-dried biosolids (B) from a municipal water treatment plant; solid pressing grape residues (G) from a spirit (pisco) producing plant; solid olive mill residues (O) from an olive oil producing plant. General properties of selected amendments are given in Table 2, and details (amendment types and doses) of experimental treatments are shown in Table 1. Application rates of organic residues were either decided according to available information in the literature for hard-rock wastes or metal-polluted degraded soils, when added alone (i.e., [40]), or with a target of 5% OM and a C:N ratio of 30 in amended tailings when added in mixtures of C-rich (i.e., grape residues, olive residues) and N-rich organic residues (i.e., biosolids, manure (i.e.,

[24])). Application rate of discarded rubble from Cu-oxide lixiviation piles was decided with a preliminary laboratory evaluation to change tailings texture from loam, to sandy loam while sediments from irrigation canals were used according to their availability. Amendments were mixed on the upper layer of tailings (0–20 cm depth) with a gasoline-operated rototiller, to avoid salinization problems in amended tailings; the exception was sediment from irrigation canals which were applied on top of tailings, as we considered them equivalent to preserved topsoils used as top covers on mine rehabilitation (i.e., [1, 13]). The experimental layout was a complete randomized block design.

Table 1: Experimental treatments and description of amendments and doses used on the field trial established at La Cocinera tailings storage facility, north-central Chile

Treatment	Amendment	Dose
Control, C	—	—
RM	Rubble from lixiviation piles	1000 $t\,ha^{-1}$
	Manure (goat)	108 $t\,ha^{-1}$
RMS	Rubble	1000 $t\,ha^{-1}$
	Manure	108 $t\,ha^{-1}$
	Sediment of irrigation canals	Layer of 5 cm
B1	Biosolids	100 $t\,ha^{-1}$
B2	Biosolids	200 $t\,ha^{-1}$
G1	Grape residues	89 $t\,ha^{-1}$
G2	Grape residues	200 $t\,ha^{-1}$
O1	Olive residues	91 $t\,ha^{-1}$
O2	Olive residues	200 $t\,ha^{-1}$
GB	Grape residues	131 $t\,ha^{-1}$
	Biosolids	20 $t\,ha^{-1}$
GM	Grape residues	91 $t\,ha^{-1}$
	Manure	70 $t\,ha^{-1}$
OB	Olive residues	135 $t\,ha^{-1}$
	Biosolids	19 $t\,ha^{-1}$
OM	Olive residues	96 $t\,ha^{-1}$
	Manure	67 $t\,ha^{-1}$

Table 2: General properties of selected organic and inorganic amendments for aided phytostabilization of Cu-sulfidic tailings under semiarid Mediterranean climate conditions

Amendment	Bulk density	pH	SOC	EC	Sulphate	Total metal (mg/kg)			Total P	Total K	Total N	Total C	C:N
	(g/mL)	(water)	(%)	(mS/cm)	(mg/L)	Cu	Zn	Fe	(mg/kg)	(mg/kg)	(%)	(%)	
Biosolids (B)	0.36	6.20	81	7.04	600	483	525	5493	9780	6638	7.4	46.9	6.3
Grape residues (G)	0.45	5.64	94	1.94	201	31	21	937	2800	24500	2.9	56.8	19.6
Olive residues (O)	0.41	5.71	95	2.91	1.25	16	11	2076	800	10500	1.0	54.3	53.8
Goat Manure (M)	0.32	7.92	71	12.18	642	46	58	8517	4100	29700	2.5	40.1	16.2
Sediments (S)	1.12	6.91	0.36	4.20	1386	333	177	n.d.	700	3400	0.12	0.02	0.2
Rubble (R)	1.70	5.28	0.23	5.55	3454	7412	510	n.d.	1300	1500	0.01	0.32	32.0

On March 2006, each experimental plot (5 × 5 m) was first amended and then seeded with 800 g of perennial ryegrass (Lolium perenne var Nui) and 38 g of a mixture of local grasses/herbs previously collected from either nearby wild areas and from a grass population (Polypogon australis) spontaneously established in a restricted area of the same TSF, following the methodology of Gold et al. [41]. This represents a seeding rate of 335 kg ha^{-1}, a value higher than the range of 20 to 120 kg ha^{-1}recommended for remediation of metal-contaminated soils of hard-rock wastes (i.e., [42–44]), but conservative considering that no information about metal tolerance of the species was available. The study area was wire-mesh fenced, and all grazing from domestic and wild herbivores was excluded; wind-breakers were built in two sides of the perimeter in order to reduce wind erosion. An irrigation system was established in the site with spray sprinklers located in the center of every experimental plot to assure plant establishment and development in the early stages. Irrigation was kept from March to November, during three consecutive years (2006 to 2008), to complement natural winter precipitations; irrigation water was obtained from a nearby stream.

Substrate Sampling and Analysis

Substrate samples were collected from all experimental plots at the time of addition of amendments (March 2006) and after two (March 2008) years, using a manual stainless steel soil auger. Each plot sample consisted of two bulked subsamples (25 cm^3 cores) randomly collected at 0–20 cm depth. A composite sample was made from mixing both subsamples which were placed in a hermetically sealed plastic bags, homogenized in the field at collection time, and stored at 4°C in the dark until their transportation to the laboratory. Roots were manually removed from all samples prior analyses and processing in the laboratory. Plot samples were divided into two aliquots; one was used for microbial analyses, and the other was used for chemical characterization as described below.

Substrate Analysis (Bulk and Pore Water)

Composite samples were made mixing replicated plots for each experimental treatment, as substrates underwent several analyses (bulk, pore water, field capacity at 100%, etc.). Substrate samples were air dried and sieved (<2 mm) before analytical determinations. The total percentage of the sample corresponding to the soil whose particle size was greater than 2 mm was registered (retained by the sieve), and the fraction less than 2 mm was determined by granulometry using the method of Bouyoucos [40]. The pH and electrical conductivity (EC) were measured in a 1:1 substrate to water solution using a glass electrode. Soil organic carbon (SOC) was determined by the Walkey and Black wet dichromate oxidation method [40]. Cation exchange capacity (CEC) and total N and Cu were determined according to protocols of the USDA [40] and US EPA [45], respectively. In the case of total Cu determinations in bulk substrate samples, every digestion batch included one blank sample, one standard reference material (SRM) sample (Loam-B, catalog CRM-LO-B; High-Purity Standard, Charleston, SC, USA), one duplicate sample, and one quality-control sample for the quality-assurance and quality-control criteria. Digested samples were analyzed for total Cu contents by flame atomic absorption spectrometry (FAAS, AAnalyst 300; Perkin-Elmer). Background nonatomic absorption was corrected with a deuterium continuous lamp. The atomic absorption analytical device was housed in a class 1000 clean-room laboratory, and the loading of the autosampler tray was done in a class 100 laminar flow cabinet. The calibration standard was prepared with high-purity (>18 $M\Omega\,cm^{-1}$) deionized water and acidified with HNO_3 Suprapur (Merck) to 0.2%. For performance control of the atomic absorption spectrometer, a certified multielement standard was used (Spectrascant Certified, Teknolab). The quality-assurance and quality-control criteria were satisfied when the measured parameter of the standard reference material (Loam-B, catalog no. CRM-LO-B; High-Purity Standard, USA) and the quality-control sample (a previously characterized soil sample with a known concentration of metals) differed by no more than 5%. Available N, P, K contents were determined according to Sadzawka et al. [46].

To assess the chemical evolution of substrates under different treatments, 35 mL of pore-water samples were taken from each experimental substrate using Rhizon soil pore-water samplers (Rhizosphere Research Products, Wageningen, The Netherlands), following the method described in Vulkan et al. [47]. Substrate pore-water samples were kept in acid-washed polyethylene plastic vials (50 mL), and then a subsample was acidified with HNO_3 suprapur

(Merck) and analyzed for total dissolved Cu (method SW-486 of US EPA [45]) by Inductively Coupled Plasma-Mass Spectrometry (ICP-MS; Perkin Elmer ELAN6100 with auto sampler). The other subsample was acidified with H_3PO_3 suprapur (Merck) and analyzed for dissolved organic carbon (DOC; method 415.1 of US EPA [48]), using an Apollo 9000 TOC analyzer (Tekmar-Dohrmann, USA).

Microbial Analysis

Substrate aliquots for microbial analysis were kept in hermetically sealed plastic bags, as collected in the field, and stored at 4°C in the dark until their microbiological analysis. Basal respiration was determined by placing 50g of each substrate sample at 70% of field capacity in a 0.5L air-tight sealed jar along with 10mL of 1N NaOH, followed by incubation for 28 days in the dark at 28°C. The C-CO_2 evolution was periodically determined by titration [49]. Basal respiration rate was calculated based on cumulative CO_2evolution over the 28 days period. Microbial biomass C (MBC) was determined by the chloroform fumigation extraction method [50]. This parameter results from the difference between fumigated and nonfumigated samples, corrected with the K_{EC} value of 0.45 [51]. The microbial metabolic quotient was calculated as basal respiration (µg C-CO_2 h^{-1}) per mg of microbial biomass C according to Anderson and Domsch [52].

Plant Sampling and Analysis

Aboveground plant biomass was determined at each experimental plot by clipping vegetation at ground level in three randomly placed quadrats (0.35 m × 0.35 m or 0.12 m^2) by the end of September 2006 and 2008, after seed production in grasses and at the mid growing season of herbs. Plant biomass of each experimental plot was harvested, placed in a preweighed paper bag, and transported to the laboratory where they were washed with tap water and deionized water in order to eliminate external contamination. Shoots were then dried in a forced air oven at 60°C and weighed after 3 d to obtain the aerial dry biomass. Plant tissues were ground to powder in an agate ball mill and digested with HNO_3–HF–H_2O_2 in a microwave oven (Milestone 1200; Milestone Microwave System, Monroe, Conn, USA). Copper content in shoots was determined by ICP-MS (Perkin Elmer ELAN 6100 with an auto sampler) according to methods SW-486 [45]. Every digestion batch included one blank sample, one SRM sample (1573a tomato leaves; National Institute of Standards and Technology, Gaithersburg, Md, USA), one duplicate sample, and one quality-control sample for the quality-assurance and quality-control criteria. The quality-assurance and quality-control criteria were satisfied when the measured parameter of the standard reference material (1573a tomato leaves) and the quality-control sample differed by no more than 5%.

Statistical Analysis

Significance of plant and microbial response variables due to experimental treatments were tested by one- (treatment) or two-way (treatment and time) analysis of variance followed by the LSD Fisher test when required. Normality and homogeneity of variances were checked with the Shapiro-Wilks and Levene tests, respectively; logarithmic transformations were used when required. Simple lineal regressions and Pearson's correlation analyses were used to determine correlations among variables (i.e., substrates). Statistical analyses were conducted using the software InfoStat [53].

RESULTS

Substrate Properties

Tailings at the experimental site have loam texture, according to the soil textural classification of the USDA (Table 3). In general, addition of amendments, either alone or in mixtures, did not modify the percentage of particle size distribution below 2,000 μm, hence maintaining the loam texture. Exceptions were RMS, O2, and OB treatments (Table 3); RMS treatment had the same sandy loam texture of pure sediments, as they were applied at the soil surface without incorporation. Slight changes in soil texture detected in the other two treatments (O2 and OB) may be only explained by typical variations of tailings texture [7]. When the coarse fraction is considered (>2,000 μm), higher percentages (among 1.4 to 3.5 times higher) are, in general, found for amended tailings when compared to control tailings (Table 3). Increases were higher for B than for G and O, and they showed to be dose dependent, as most of the OM incorporated into experimental plots was retained on the sieve. As expected, the highest values of coarse particle fraction (>2,000 μm) were found on plots where rubble from Cu-lixiviation piles was incorporated (Table 3), as this material has 68% of particles >2,000 μm.

Table 3: Texture of experimental substrates (bulk) at the time of establishment of the field assay (year 1, 2006), at a depth of 0–20 cm. Codes of treatments follow Table 1

Treatment	Texture (%)			Texture type*	Particles >2,000 μm
	<2 μm	2 μm – 50 μm	50 μm – 2,000 μm		(%)
C	14	42	44	Loam	11
RM	13	33	54	Loam	39
RMS	12	25	63	Sandy loam	39
B1	12	43	45	Loam	21
B2	20	43	37	Loam	38
G1	16	44	40	Loam	15
G2	19	40	41	Loam	19
O1	9	37	54	Loam	8
O2	11	30	59	Sandy Loam	18
GB	21	42	37	Loam	15
GM	15	37	47	Loam	20
OB	11	50	39	Silt Loam	19
OM	18	40	42	Loam	17

Footnotes

*According to the soil textural classification chart of the US Department of Agriculture.

At the beginning of the assay (year 1 or 2006), addition of amendments improved most chemical (i.e., CEC, SOC, and DOC) and nutritional (i.e., N, P) properties of tailings (Tables 3 and 4). Biosolids and grape residues, either alone or mixed with other materials, produced the more marked changes in chemical and nutritional properties of tailings, particularly in terms of CEC, SOC, and available N-P-K concentrations, when compared to control (Table 4). Olive residues mainly improved SOC of tailings in a dose-dependent form (Table 4). However, all organic amendments contained soluble salts, and their addition into tailings increased EC with respect to control ($5.3\,mS\,cm^{-1}$), up to $26.4\,mS\,cm^{-1}$, particularly when applied in high doses; this effect was more pronounced with biosolids (Table 4) but also occurred after addition of high dosages ($200\,t\,ha^{-1}$) of grape and olive residues.

The carbon to nitrogen ratio largely varied among treatments and generally decreased with addition of organic amendments, particularly in the case of biosolids and manure as these are N-rich materials (Table 4). Incorporation of amendments into tailings did not change or slightly reduced total Cu concentrations in the substrate, thus having no or minor dilution effects. The exception was treatment G2 with addition of $200\,t\,ha^{-1}$ of grape residues which reduced 1.8 times total Cu (Table 4). The pH of substrates only varied from slightly too moderately alkaline values, in the range of 7.40 to 8.20, and it almost did not vary with time (Table 4).

Table 4: Evolution of chemical properties of experimental substrates (bulk) from the time of establishment of the field assay (year 1, 2006) and after 2 years (year 3, 2008), at a depth of 0–20 cm. Codes of treatments follow Table 1

Treat-ment	pH		EC (mS cm^{-1})		CEC (meq 100 g^{-1})		SOC (%)		$N_{available}$ (mg kg^{-1})		$P_{available}$ (mg kg^{-1})		$K_{available}$ (mg kg^{-1})		C:N ratio		Total Cu (mg kg^{-1})	
	Y 1	Y3	Y 1	Y3	Y 1	Y3	Y 1	Y3	Y 1	Y3	Y 1	Y3	Y 1	Y3	Y 1	Y3	Y 1	Y3
C	8.0	8.0	5.3	7.4	3,7	5.8	2.0	0.7	9	Bdl	8	8	142	73	56.6	6.7	6248	6140
RM	7.6	7.9	9.1	11.9	18,1	7.8	3.3	1.5	14	13	139	53	2821	826	12.7	7.9	6445	6604
RMS	7.4	7.8	5.9	6.9	21,5	9.7	2.6	1.2	35	11	42	25	469	269	18.7	9.0	4910	5728
B1	8.1	8.0	13.5	15.0	9,3	4.9	2.0	1.1	480	425	287	47	616	232	10.8	4.3	4655	5876
B2	7.8	7.7	26.4	19.6	31,5	7.0	5.0	1.3	984	814	993	87	2127	404	6.4	4.9	4013	4839
G1	8.2	7.9	8.9	7.5	12,6	4.9	2.5	1.4	24	9	54	13	1264	155	48.3	9.1	6506	6879
G2	7.6	8.0	12.0	8.9	24,0	8.2	3.4	2.0	180	6	269	40	4727	422	8.5	7.3	3394	5208
O1	8.1	8.0	8.5	6.5	6,5	5.8	2.4	1.1	5	Bdl	15	8	713	81	28.3	14.4	8588	7369
O2	7.8	8.0	10.2	7.9	8,4	5.6	5.8	1.6	26	Bdl	16	8	1464	188	37.3	12.8	6739	5215
GB	8.0	8.1	12.0	9.3	15,5	4.5	3.3	1.3	444	32	269	41	1850	295	21.2	9.1	7163	5928
GM	8.2	8.0	13.1	8.7	23,6	6.3	4.8	1.7	207	22	175	33	3088	333	91.8	7.4	7391	7353
OB	7.8	8.0	9.7	6.6	7,3	10.0	4.4	1.5	178	Bdl	77	16	885	102	23.4	12.6	5856	6178
OM	8.0	8.2	14.5	10.8	36,5	8.0	5.1	1.1	24	Bdl	84	20	3055	512	19.9	6.4	5857	5317

Y1: year 1; Y3: year 3; EC: electrical conductivity; CEC: cation exchange capacity; SOC: soil organic carbon; Bld: below detection limit.

By the end of the experiment (year 3 or 2008), CEC, SOC, available N-P-K, and C:N ratio markedly decreased in all treatments, but in all cases values were higher than control plots (Table 4). Even though available N tended to decrease with time, minor reductions were detected on biosolids-amended plots when compared to other organic amendments (Table 4). Total Cu concentrations in the substrate did not show relevant variations with time while EC values either slightly decreased or increased with time (Table 4).

Dissolved organic carbon in pore water was very low in tailings (control plots), but it increased by 1 to 2 orders of magnitude with incorporation of amendments (Table 5). Specifically, DOC reached highest values in those plots where only biosolids were added (B1 and B2). In general, DOC values decreased with time, but they were more marked in some treatments, such as B1 and B2 (Table 5). Total dissolved Cu in pore water of substrates increased from 17 to 2,923 times with addition of amendments when compared to control plots (Table 5). Biosolids amended plots showed the higher increases reaching concentrations of 29.1 and 116.9 mg L^{-1} in treatments B1 and B2, respectively (Table 5). However, Cu concentrations in pore water markedly decreased with time, reaching values quite alike among treatments after three years (Table 5). During the first year of the study, a positive and significant relation between DOC and total dissolved Cu in pore water was found (R^2=0.75, P<0.05), but by the third year this relation was not significant (R^2=0.01, P<0.73).

Table 5: Evolution of the concentration of dissolved organic carbon (DOC) and total dissolved Cu in the pore water of substrates from the time of establishment of the field assay (year 1 or 2006) and after 2 years (year 3 or 2008). Codes of treatments follow Table 1

Treatment	DOC (mg L^{-1})		Cu (mg L^{-1})	
	Y1	Y3	Y1	Y3
C	11	29	0.04	0.17
RM	542	795	4.52	0.06
RMS	218	148	1.22	4.01
B1	2412	274	29.12	0.57
B2	3048	600	116.92	1.13
G1	202	72	2.81	0.11

G2	523	113	8.24	0.05
O1	310	57	0.69	0.65
O2	831	138	2.28	0.51
GB	516	253	5.63	1.24
GM	780	190	7.79	1.00
OB	813	243	28.04	0.40
OM	918	165	6.96	2.44

Footnotes

Y1: year 1; Y3: year 3.

Plant Responses

Figure 1 shows the variation of aerial plant biomass (dry weight basis, d.w.) among experimental treatments with time, while Table 6 shows variation of Cu content in shoots. A two-way ANOVA for aerial plant biomass indicated significant differences among experimental treatments ($F=7.08$, $P<0.05$), year since amendments addition ($F=51.07$, $P<0.05$), and the interaction among these factors ($F=7.46$, $P<0.05$). Control plots showed very low aerial biomass production (4.5 to 5 g m^{-2}dry weight basis (d.w.)), irrespective of the year (Figure 1). During the first growing season, aerial plant biomass strongly varied among treatments, being higher on treatments RMS and G1 (19 and 11 g m^{-2}d.w., resp.) and null or very limited on treatments with addition of biosolids (0 to 0.3 g m^{-2} d.w.; Figure1); all other treatments showed an aerial biomass production that ranged from 2.5 to 7.8 g m^{-2} d.w. Plant biomass tended to increase with time on most treatments (Figure 1). However, after three years, the highest increase in aerial plant biomass production occurred on treatment B1 (480 g m^{-2}, d.w.), followed by treatments B2 (283 g m^{-2} d.w.) and RM (112 g m^{-2} d.w.). On the contrary, treatments O1 and O2, with addition of olive residues had marked reductions in aerial production, reaching values of only 0.1 to 0.2 g m^{-2} d.w. (Figure 1).

Table 6: Evolution of the concentration of copper in aerial plant biomass (mean ± standard deviation, n = 3) from the time of establishment of the field assay (year 1 or 2006) and after 2 years (year 3 or 2008). Codes of treatments follow Table 1. Values followed by the same letter are not significantly different at P < 0.05 according to a two-way ANOVA and LSD Fisher test

Treatment	Cu in shoot (mg kg^{-1})	
	Y1	Y3
C	237 ± 120.3^{abc}	203 ± 19.6^{abc}
RM	310 ± 215.7^{bcd}	390 ±433.4^{cde}
RMS	90 ± 8.3^{a}	270 ± 217.0^{bcd}
B1	826 ± 480.6^{f}	207 ± 90.5^{abc}
B2	—	144 ± 40.9^{ab}
G1	189 ± 69.8^{abc}	345 ± 268.8^{bcde}
G2	231 ± 114.9^{abc}	187 ± 48.4^{abc}
O1	559 ± 428.9^{e}	—
O2	507 ± 493.2^{de}	280 ± 202.4^{bc}
GB	427 ± 274.4^{de}	154 ± 31.5^{ab}
GM	316 ± 293.3^{bcde}	180 ± 35.4^{abc}
OB	235 ± 104.9^{abc}	207 ± 94.2^{abc}
OM	275 ± 83.4^{bcd}	312 ± 162.8^{bcde}

Footnotes

Y1: year 1; Y3: year 3.

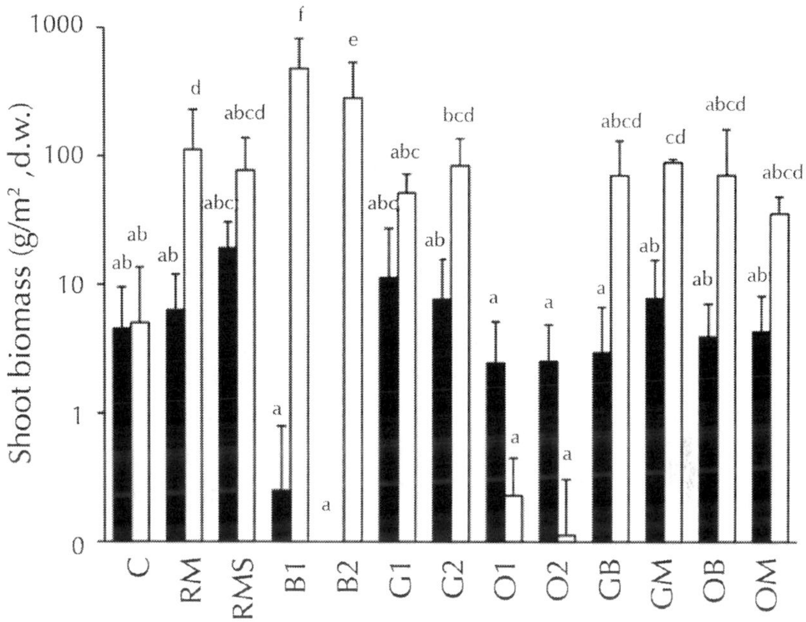

Figure 1: Evolution of plant aerial biomass (dry weight basis: d.w, log 10 scale) in experimental treatments (mean ± standard deviation, n=3) from the time of establishment of the field assay (year 1 or 2006 in black columns) and after 2 years (year 3 or 2008 in white columns). Codes of treatments follow Table 1. Same letters indicate no significant differences at P<0.05 according to a two-way ANOVA and LSD Fisher test.

A two-way ANOVA for Cu concentration in shoots indicated significant differences among experimental treatments (F=2.33, P<0.05), year since amendments addition (F=7.34, P<0.05), and the interaction among these factors (F=3.52, P<0.05). Plants in control plots reached mean values of 203–237 mg Cu kg^{-1} in aerial tissues with no significant variation among years (Table 6). During the first year, all other treatments did not show significant differences in Cu concentrations in shoots (i.e., RM, RMS, G1, G2, OB, OM) or showed significant increases (B1, O1, O2, GB) when compared to control plots. The maximum Cu concentrations in shoots occurred in treatment B1 (826 mg Cu kg^{-1}), followed by treatments O1, O2, and GB that showed values that ranged from 427 to 559 mg kg^{-1}. In most treatments, Cu contents in shoots remained the same or significantly decreased with time (Table 6). Specifically, Cu contents in shoots of plants in treatments B1, O2, and GB were reduced from 2 to 4 times with time (Table 6). With the exception of control and treatments with olive residues addition (O1 and O2), no visual metal toxicity symptoms (i.e., redness and chlorotic leaves, stunted plants) were, however, detected on field plants. Furthermore, biomass production increased with time in most treatments as shown above (Figure 1), with exception of control and treatments with olive residues, thus indicating no metal phytotoxicity of most amended substrates.

Microbiological Properties

Results of microbiological parameters are given in Figure 2 (accumulated basal respiration), Figure 3(biomass C), and Table 7 (metabolic quotient). Values of both microbial basal respiration (MBR; Figure2) and microbial biomass C (MBC; Figure 3) were low and constant in time in control plots, while the metabolic quotient (Table 7) was higher and constant in time in control plots when compared to experimental treatments.

Table 7: Evolution of metabolic quotient (mean ± standard deviation, n = 3) in experimental substrates (0-20 cm) from the time of establishment of the field assay (year 1 or 2006) and after 2 years (year 3 or 2008). Codes of treatments follow Table1. Values in the same column followed by the same letter are not significantly different at P<0.05 according to ANOVA

Treatment	Metabolic quotient (μg C-CO_2 mg Cbio^{-1} h^{-1})	
	Y1	Y3
C	5.0 ± 0.75^{a}	4.0 ± 0.83^{a}
RM	3.0 ± 0.50^{b}	2.0 ± 0.62^{b}
RMS	3.0 ± 0.35^{b}	1.6 ± 0.71^{b}
B1	6.9 ± 0.68^{c}	0.6 ± 0.51^{c}
B2	7.5 ± 0.71^{d}	0.9 ± 0.77^{c}
G1	3.2 ± 0.51^{b}	2.8 ± 0.68^{d}
G2	3.4 ± 0.65^{b}	1.2 ± 0.80^{bc}
O1	7.1 ± 0.92^{c}	4.7 ± 1.03^{a}
O2	7.5 ± 1.09^{d}	4.8 ± 0.92^{a}
GB	3.9 ± 0.71^{bc}	1.6 ± 0.73^{b}

GM	3.7 ± 0.78[bc]	1.2 ± 0.50[bc]
OB	3.6 ± 0.80[b]	1.8 ± 0.62[b]
OM	4.7 ± 0.85[ab]	2.9 ± 0.75[b]

Footnotes

Y1: year 1; Y3: year 3; Cbio: microbial biomass C.

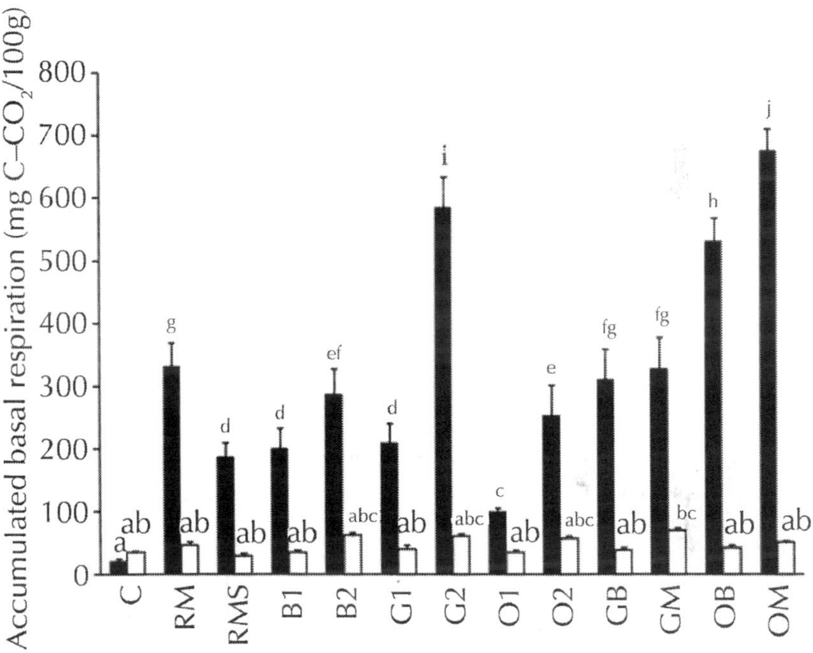

Figure 2: Evolution of accumulated basal respiration (28 days; mean ± standard deviation, n=3) in experimental substrates from the time of establishment of the field assay (year 1 or 2006 in black columns) and after 2 years (year 3 or 2008 in white columns). Codes of treatments follow Table 1. Same letters indicate no significant differences at $P<0.05$ according to a two-way ANOVA and LSD Fisher test.

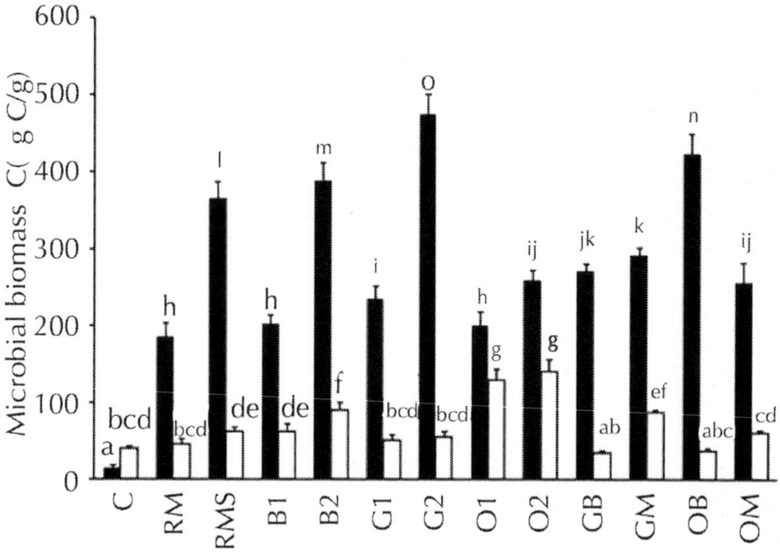

Figure 3: Evolution of microbial biomass carbon (MBC; mean ± standard deviation, n=3) in experimental substrates from the time of establishment of the field assay (year 1 or 2006 in black columns) and after 2 years (year 3 or 2008 in white columns). Codes of treatments follow Table 1. Same letters indicate no significant differences at $P<0.05$ according to a two-way ANOVA and LSD Fisher test.

Two-way ANOVAs for MBR and MBC indicated in both cases significant differences among experimental treatments, year since amendments addition, and the interaction among these factors (Table 8). At the beginning of the assay (year 1 or 2006), both MBR and MBC significantly increased with addition of all amendments (Figures 2 and 3). Increase in MBR was higher in treatments RM, G2, OB, and OM (from 16 to 32 times) and lower in treatment O1 (5 times) with respect to control. In the case of MBC, increase was higher in treatments G2, OB, B2, and RMS (from 24 to 32 times) with respect to control; furthermore, for both parameters and all organic amendments used (B, G, O), the increase was dose dependent. MBR and MBC tended to decreased with time in all treatments, MBR showing more marked reductions than MBC (Figures 2 and 3). After three years, BMR values of all treatments reached values similar to control plots, with the exception of treatment GM that showed values 2 times higher than control plots (Figure 2).

In the case of MBC, even though this parameter decreased in time to values similar to control plots, treatments B2, O1, and O2 remained significantly higher than control plots, reaching values up to 3.5 times higher than control (Figure 3). A significant simple lineal regression existed among MBR and MBC (R^2=0.37, F=21.7, P<0.05) during the first year of the assay, but this disappeared after three years (R2=0.09, F=3.8, P=0.0577) of experimentation. Finally, the metabolic quotient tended to decrease with time (Table 7). The highest values were recorded in treatments O1 and O2 for both years of evaluation. Treatments B1 and B2 showed high values in year 1, similar to treatments O1 and O2, but markedly decreased on the following years, reaching the lowest values in year 3 (Table 7).

Table 8: Tables of two-way analysis of variance for microbial accumulated basal respiration and microbial biomass C (mean ± standard deviation,n=3) in experimental substrates (0–20 cm) from the time of establishment of the field assay (year 1 or 2006) and after 2 years (year 3 or 2008)

Parameter	Source of variation	Degrees of freedom	Mean squares	F	P
Basal respiration	Model	25	105727	152.2	<0.01
	Treatment	12	57253	82.4	<0.01
	Year	1	1352296	1947.2	<0.01
	Treatment × Year	12	50320	72.5	<0.01
	Error	52	695		
Biomass C	Model	25	55094	253.9	<0.01
	Treatment	12	23278	107.3	<0.01
	Year	1	819816	3779.4	<0.01
	Treatment × Year	12	23183	106.9	<0.01
	Error	52	217		

Correlation between Biological and Chemical Parameters of Substrates

Pearson's correlations between all soil properties (bulk and pore water) and plant-related parameters were calculated. The most relevant are shown in Table 9. The strength of the associations was interpreted according to the Hopkins' correlation classification [54]: insubstantial (0.0–0.1), low (0.1–0.3), moderate (0.3–0.5), high (0.5–0.7), very high (0.7–0.9), and nearly perfect (0.9–1.0). Correlation analysis of plant aerial biomass showed high positive relationship with SOC, MBR, and Cu concentration in pore water; a very high negative relationship with metabolic quotient; a very high positive relationship with MBC; a moderate negative relationship with Cu concentration in shoots.

Table 9: Pearson's correlation coefficients between plant aerial biomass and several microbial and chemical (bulk and pore water) properties of experimental substrates (n=39)

Variable	Plant aerial biomass
DOC	−0.20
SOC	0.54
EC	0.47
Microbial basal respiration	0.55
Metabolic quotient	−0.65
Cu concentration in pore water	0.51
Microbial biomass C	0.60
Cu concentration in shoots	−0.42

DOC: dissolved organic carbon; SOC: soil organic carbon; EC: electric conductivity.

DISCUSSION

Substrates Properties

Soils are generally classified as saline when they have EC values higher than 4 mS cm^{-1} [55]. Under this guideline, all treatments were moderately to strongly saline, even after three years since establishment, and even though organic amendments were incorporated into tailings (mixed).

Secondary salinisation of substrates, due to application of organic residues, can be a major constraint in aided phytostabilization strategies, particularly under semiarid environments where evapotranspiration exceeds infiltration of water due to scarce rain events (e.g., [56]). For example, Lombard et al. [37] found that a composted biosolid-treated calcareous soil increased EC by 21 and 41% above the control soil when applied at rates of only 14 t ha^{-1} and 44.5 t ha^{-1}, respectively, in an area with semiarid climate. The increase in EC was expected because biosolids are high in complex mixtures of soluble salts [37]. Salinisation of tailings (increased EC) by amendments may be limiting for plant establishment and growth, but it would be compensated by high CEC of organic amendments (with the exception of olive residues) and discarded rubble from lixiviation piles. In semiarid climates, salinity is usually combined with high soil pH (alkalinity), because of $CaCO_3$ enrichment in the uppermost soil layers; hence calcareous soils are common; in the case of the experimental area, pH of substrates only ranged from slightly to moderately alkaline, but Ca content of tailings is high (1.5%). Although the effect of pH on the activity of soil microorganisms has been reported [57, 58], variations in soil pH in the present study do not seem to be sufficiently great to establish differences in the biological activity since no correlation between MBC and pH was found.

The average SOC in mine tailings of semiarid areas of north-central Chile is only about 0.23% [39]. This might predict low or lack of microbiological activity. Indeed, indigenous microbial communities associated with mine tailings often show limited density and diversity, relative to undisturbed sites [11,59], and mainly correspond to iron-/sulfur-oxidizing bacteria [12]. However, SOC reached a value of 2% in La Cocinera TSF. Most of the SOC in tailings probably comes from flotation reagents used for the copper concentration process in the flotation plant. The level of SOC of all organically amended tailings increased noticeably with respect to control during the first year, in agreement with literature (i.e., [31]), but it decreased along the third year of the experiment, probably due to mineralization of easily degradable materials added [60]; however, in all cases, it remained above that of control plots and thus assuring microbial activity and plant growth. This may be also attributed to the presence of the plant cover developed after amendments application, with organic inputs such as root exudates and plant remains compensating C losses through mineralization, and due to the exhaustion of the labile organic C and the increase of the recalcitrant fraction of C, which is less prone to mineralization [61, 62]. For example, Müller da Silva et al. [27] found beneficial changes in ecosystem functioning, such as enhanced biological cycling of nutrients, in Eucalyptus plantations where biosolids were incorporated into degraded soils; as a result of increased plant biomass production, due to improved OM and nutrient soil levels, higher transfers of these components to the litter and subsequently to the soil were measured; if litter degradation is adequate, then self-sustainability of the system may be reached with no need of further additions of amendments, particularly for plant formations adapted to nutrient-poor soils like in semiarid areas. Furthermore, the slow release of nutrients contained within the biosolids and other organic amendments, when compared to chemical fertilizers [63], makes it possible to restore soil nutrient stocks throughout the development cycle of revegetated areas. Longer-term evaluations of the experimental plots established in the present study, in terms of OM and nutrients cycling among other parameters, are thus required for better conclusions.

Addition of organic amendments decreased C:N ratio of tailings, hence improving fertility of substrate; however, a large variation in this parameter existed among treatments, as locally available organic amendments have different N concentrations. For example, biosolids have large N concentrations (7.4%), while grape and olive residues have much lower concentrations (2.9% and 1%, resp.). Variations in C:N ratio determine different effects on microbiological processes and therefore in N availability to plants, not only in the short term but also in the long term (i.e., [24, 64]). For example, the decrease of this ratio to recommended values of 20 to 30 means that N is mineralized and thus available for soil microorganisms and plant roots [65]. However, very low (<20) C:N ratios may generate excessive N mineralization with consequent lixiviation of nitrates to groundwater and release of ammonia [27, 66] and CO_2 to the atmosphere [67], thus restricting seed germination and early stages of plant development [27, 66]. This effect may have occurred on biosolids-amended tailings, particularly when elevate doses were applied (treatment B2), as no seed germination was observed during the first year (data not given). On the other hand, elevated values of this ratio (>30) indicate low N concentrations, which is a less favorable condition for soil microbial activity and plant development; this last situation is further limiting if the main C forms of the material are recalcitrant (i.e., lignin), which is resistant to microbial degradation [68]. In this case, N is sequestered into soil microorganisms and not available for plant root uptake [69]. For example, addition of olive residues resulted in elevated C:N rates (>28), and it remained high after three years. A shortage of N, due to the low content of this element in olive residues, may have occurred after three years thus explaining the marked reduction in aerial biomass production under these treatments (O1 and O2); mixture of olive residues with other N-rich materials, such as biosolids and goat manure, allowed longer-term N availability, and thus sustained aerial plant biomass production in time should be possible without further additions to maintain adequate fertility.

The organic amendments increased the initial level of DOC, an effect which was still present at the third year but with much reduced concentrations. The high content of low molecular weight organic acids in biosolids and olive residues at dressing time, such as fulvic acids, may pose a metal toxicity risk for plants due to their capability for increasing metal solubility and bioavailability in the substrate solution (e.g., [31, 32, 70, 71]). Metal cations, such as Cu^{2+}, are complexed by DOC, and these soluble organo-metallic complexes can be readily absorbed by plant roots [31, 32, 72–75] and/or leached into deep substrate layers [76]. Indeed, higher copper contents have been detected in shoots of grasses and some trees growing on biosolids-mixed tailings [25] and soils [31, 32, and 77]. This phenomenon was detected in the present study in both biosolids and olive residues amended tailings, particularly during the first year. A decrease in the concentration of soluble metals would be expected on the longer term, as detected in the present study, once the mineralization of labile organic matter in biosolids or other organic materials leads to stabilized organic matter, as it has been shown by Al-Wabel et al. [78].

Biological Properties

The establishment of a vegetation cover on mine tailings located in semiarid Mediterranean climates is fundamental to protect these sites against erosive processes and for in situ immobilization of metals in the substrate; at the same time, it contributes to increase the soil organic matter content. In addition, a plant cover and particularly, the builtup of plant rhizosphere will influence the biological quality of the substrate by favoring the soil microbial activity and phytostabilization of these sites. On the other hand, soil microbiota is a key component to assure nutrient cycling for plant availability and long-term sustainability.

Application of amendments on mine tailings under semiarid Mediterranean climatic conditions improved the biological properties of original substrate. Organic amendments also favored plant growth which protect substrate from natural forces (i.e., wind and rain) and contribute to phytostabilization. The species used in the present study were able to survive in the tailings, but growth measured as dry aerial biomass largely varied among treatments. Biosolids inhibited seed germination (data not shown) and showed to be limiting for plant establishment and growth at the beginning of the assay, irrespective of application dose (100 and 200 $t\,ha^{-1}$). This may be result of the high salinity of this material, but also due to the large concentration of DOC which mobilized Cu into soil pore water thus posing Cu toxicity risks to plants. However, at the end of the assay, these treatments showed the highest plant yields, maybe as a result of salt and DOC leaching through the profile due to irrigation and natural precipitation, as it has been shown in other field studies where biosolids were used as organic amendments for mine tailings (e.g., [25, 56]). Furthermore, temporal changes in C:N ratio of organic-amended tailings may also affect plant establishment and productivity as demonstrated by Brown et al. [64]. They demonstrated that increasing the C:N ratio of organic amendments added to mine tailings to ≥20:1 increased plant species richness and growth, thus affecting native plant restoration.

Olive residues allowed plant establishment but they were inadequate for sustained plant growth, showing poor yield of biomass after three years; however, they improved microbial properties (MSR and MBC) of tailings even after three years since plot establishment. One explanation of this result may be based on the presence of phenolic compounds, which are toxic [79, 80]. The organic matter of olive residues mainly consists of oil, polysaccharides, sugars, polyphenols, polyalcohols, proteins, organic acids (i.e., acetic and formic acid), phenols, lipids, and tannins, some of them known to be toxic to plants [70] or biorecalcitrant [79, 81]. However, inorganic constituents at the concentration levels found in olive residues are not toxic. In fact, it has been proved that they may potentially act as a good source of plant nutrients [82–84]. According to this, the use of composted olive residues is suggested as when olive residues undergo through a proper biodegradation process, like composting, the toxic organic compounds are broken down, and remaining components of these residues are suitable as good source of plant and microbial nutrients [84]. Other explanation may be the low N fertility of this material as discussed above, as when olive residues were mixed with biosolids and goat manure (N-rich sources), plant biomass increased 59% with respect to control. Thus it is not advisable to apply directly olive residues to mine tailings for phytostabilization. Several treatments showed to be adequate for plant establishment and growth in the short term, besides of improving microbiological properties of tailings (e.g., RM, B1, B2, OB). These treatments showed the highest biomass yield, MSR, and MBC; however, plots have to be evaluated in the longer term for better conclusions.

The negative correlation between qCO_2 and plant aerial biomass found in the present study reflects environmental stress probably as a result of phytotoxicity effects from some organic amendments mentioned above, like olive residues and biosolids. Treatments which had the higher plant biomass yield showed the lower qCO_2 values. On the other hand, a positive correlation was found between MBC and plant aerial biomass, confirming the usefulness of MBC as an indicator of changes in vegetation. These results suggest that qCO_2 and MBC constitute sensitive indicators of plant growth and thus for phytostabilization progress [85]. The present study showed that herbaceous/grass species can grow in a wide range of metal concentrations (Cu, Zn, and Fe) in mine tailings. Zinc and Cu are essential for normal plant growth and development at low concentrations [86] and play important roles in several metabolic processes in plants. However, excess Cu and Zn in soil may retard plant growth [87, 88]. Kabata-Pendias and Pendias [89] reported that total fractions in soil equal to 70 to 400 mg kg^{-1} of Zn and 60 to 125 mg kg^{-1} of Cu are toxic to plants. The metal contents in the substrates studied greatly exceeded these ranges.

With regard to metal accumulation in shoots, it has been stated in the literature that organic amendments that contain a high proportion of humified OM (i.e., compost) can decrease the mobility of some metals due to the formation of stable chelates [90, 91]. However, the residues used in the present study were not stabilized at the beginning of the assay, and the humified OM could have been low, as shown by elevated DOC levels in pore water. Instead, they might facilitate metal transport in substrate by acting as carriers through formation of metal-organic complexes [92], at least in the short term. As a result, Cu contents in shoots were alike or slightly higher than in control plots. However, the concentration of metals in shoots tended to decrease with time in most treatments, suggesting that an immobilization process is taking place, probably because of organic matter (OM) stabilization. Shoot copper concentrations found in all treatments were well above critical concentrations described in plants and the maximum tolerable level in animals [88, 89, and 93]. Nevertheless, it is important to state some relevant aspects. First, critical concentrations of copper have been defined for sensitive plants, normally represented by crops and vegetables (i.e., [88, 89, 93]); in the present study, local adapted plants were used, which may be more tolerant to elevated copper levels in aerial tissues. Second, some researchers have reported that part of the metals found in shoots might be adhered onto stem and leaves surface and are not absorbed into the internal structures of the plant, suggesting that total metal content in shoots is overestimated (e.g. [87, 94, 95]). This external contamination may exist even after standard washing protocols, particularly in xerophytic plants of semiarid environments as trichomes and glands are common morphological structures on leaves adapted to drought (i.e., [96]). Finally, the maximum tolerable level of copper in animals assumes that animal has no other feed or forage source, which is not the case, because these sites are not destined for grazing. Thus, the high concentration of copper in shoots reported in the present study does not necessarily imply risk for food webs.

CONCLUSIONS

Aided phytostabilization based on the use of local amendments and plant sources is a feasible field-scale technology for large-scale postoperative Cu-sulfidic mine TSF under semiarid Mediterranean climate conditions. A broad range of organic- and hard-rock mine wastes showed to be adequate for improving chemical and biological properties of tailings. However, selection of adequate local available amendments and management options for long-term release of limiting conditions of tailings for microbial and plant establishment and development are key aspects.

Biosolids, grape residues, and goat manure, either alone or in mixtures, are adequate organic materials for improving chemical and biological properties of tailings; however, high doses should be avoided (≥200 ton ha^{-1} d.w.), particularly in the case of biosolids, as salinization, ammonia volatilization, and Cu mobilization into pore water may secondary occur with consequent limiting conditions for plant establishment. Even though biosolids are useful organic amendments to speed up the builtup of plant biomass and restore microbial properties of Cu-sulfidic mine tailings, lixiviation of excess salts and DOC during early stages of aided phytostabilization is required in order to get proper plant establishment and growth.

Olive residues are adequate materials for restoring microbial properties of tailings, but their low N fertility and high content of phytotoxic compounds made it inadequate to sustain plant development in the long term. Their mixture with N-rich materials, such as biosolids and manure, would be a better management option for these applications; furthermore, preliminary composting may be another alternative, particularly to eliminate the compounds responsible of phytotoxicity, but this needs to be further evaluated.

Discarded rubble from copper-oxide lixiviation piles can be a useful amendment for tailings, but it needs to be mixed with organic amendments (i.e., manure) to assure microbial inoculation of tailings and improvement of nutritional properties that may limit plant establishment and grow. Furthermore, general chemical characteristics of rubble and tailings where it will be incorporated should be first evaluated in order to determine its efficacy and define specific management options (i.e., pH management).

Local grass/herb species are appropriate for aided phytostabilization of abandoned and postoperative TSF under semiarid Mediterranean climate conditions, as they rapidly build up a continuous plant cover, but high Cu concentrations found in shoots may increase the potential risk of metal transfer to the food chain, an aspect that should be further evaluated, considering background metal contents in wild plants.

ACKNOWLEDGMENTS

The present study was funded by the Innova Chile CORFO project 04CR9IXD-01 to R. Ginocchio. The authors would like to acknowledge the permanent support of Mr. Marcos Velásquez, mining engineer of Planta Ovalle, ENAMI, Ismael Jiménez and Jaime Cuevas of INIA Intihuasi with the establishment and maintenance of the field assay.

REFERENCES

1. A. D. Bradshaw, "The reconstruction of ecosystems," Journal of Applied Ecology, vol. 20, no. 1, pp. 1–17, 1983.
2. J. McCall, J. Gunn, and H. Struik, "Photo interpretive study of recovery of damaged lands near the metal smelters of Sudbury, Canada," Water, Air, and Soil Pollution, vol. 85, no. 2, pp. 847–852, 1995.
3. R. Badilla-Ohlbaum, R. Ginocchio, P. H. Rodríguez et al., "Relationship between soil copper content and copper content of selected crop plants in central Chile," Environmental Toxicology and Chemistry, vol. 20, no. 12, pp. 2749–2757, 2001.
4. R. Ginocchio, P. Sánchez, L. M. de la Fuente et al., "Agricultural soils spiked with copper mine wastes and copper concentrate: implications for copper bioavailability and bioaccumulation,"Environmental Toxicology and Chemistry, vol. 25, no. 3, pp. 712–718, 2006.
5. M. O. Mendez and R. M. Maier, "Phytoremediation of mine tailings in temperate and arid environments," Reviews in Environmental Science and Biotechnology, vol. 7, no. 1, pp. 47–59, 2008.
6. R. Ginocchio, L. M. de la Fuente, P. Sánchez et al., "Soil acidification as a confounding factor on metal phytotoxicity in soils spiked with copper-rich mine wastes," Environmental Toxicology and Chemistry, vol. 28, no. 10, pp. 2069–2081, 2009.
7. B. Dold and L. Fontboté, "Element cycling and secondary mineralogy in porphyry copper tailings as a function of climate, primary mineralogy, and mineral processing," Journal of Geochemical Exploration, vol. 74, no. 1–3, pp. 3–55, 2001.
8. R. Ginocchio, "Effects of a copper smelter on a grassland community in the Puchuncaví Valley, Chile," Chemosphere, vol. 41, no. 1-2, pp. 15–23, 2000.
9. R. Ginocchio, P. H. Rodríguez, R. Badilla-Ohlbaum, H. E. Allen, and G. E. Lagos, "Effect of soil copper content and pH on copper uptake of selected vegetables grown under controlled

conditions," Environmental Toxicology and Chemistry, vol. 21, no. 8, pp. 1736–1744, 2002.

10. I. G. Petrisor, D. Dobrota, K. Komnitsas, I. Lazar, J. M. Kuperberg, and M. Serban, "Artificial inoculation—perspectives in tailings phytostabilization," International Journal of Phytoremediation, vol. 1, no. 1, pp. 1–15, 2004.
11. R. de la Iglesia, D. Castro, R. Ginocchio, D. van der Lelie, and B. González, "Factors influencing the composition of bacterial communities found at abandoned copper-tailings dumps," Journal of Applied Microbiology, vol. 100, no. 3, pp. 537–544, 2006.
12. N. Diaby, B. Dold, H. R. Pfeifer, C. Holliger, D. B. Johnson, and K. B. Hallberg, "Microbial communities in a porphyry copper tailings impoundment and their impact on the geochemical dynamics of the mine waste," Environmental Microbiology, vol. 9, no. 2, pp. 298–307, 2007.
13. F. F. Munshower, Practical Handbook of Disturbed Land Revegetation, Lewis Publisher, Boca Raton, Fla, USA, 1993.
14. W. Krzaklewski and M. Pietrzykowski, "Selected physico-chemical properties of zinc and lead ore tailings and their biological stabilisation," Water, Air, and Soil Pollution, vol. 141, no. 1–4, pp. 125–142, 2002.
15. R. Ginocchio, C. Santibáñez, P. León-Lobos, S. Brown, and A. J. M. Baker, "Sustainable rehabilitation of copper mine tailings in Chile through phytostabilization: more than plants," in Proceedings of the 2nd International Seminar on Mine Closure, A. Fourie, M. Tibbet, and J. Wiertz, Eds., pp. 465–474, Australian Centre for Geomechanics, Santiago, Chile, 2007.
16. Z. H. Ye, Z. Y. Yang, G. Y. S. Chan, and M. H. Wong, "Growth response of Sesbania rostrata andS. cannabina to sludge-amended lead/zinc mine tailings. A greenhouse study," Environment International, vol. 26, no. 5-6, pp. 449–455, 2001.
17. Z. Y. Yang, J. G. Yuan, G. R. Xin, H. T. Chang, and M. H. Wong, "Germination, growth, and nodulation of Sesbania rostrata grown in Pb/Zn mine tailings," Environmental Management, vol. 21, no. 4, pp. 617–622, 1997.
18. R. R. Brooks, "Geobotany and hyperaccumulators," in Plants that Hyperaccumulate Heavy Metals: Their Role in Phytoremediation,

Microbiology, Archaeology, Mineral Exploration, and Phytomining, R. R. Brooks, Ed., pp. 55–94, CAB International, Wallingford, UK, 1998.

19. J. Vangronsveld, R. Herzig, N. Weyens et al., "Phytoremediation of contaminated soils and groundwater: lessons from the field," Environmental Science & Pollution Research, vol. 16, no. 7, pp. 765–794, 2009.
20. M. Mench, N. Lepp, V. Bert et al., "Successes and limitations of phytotechnologies at field scale: outcomes, assessment and outlook from COST action 859," Journal of Soils and Sediments, vol. 10, no. 6, pp. 1039–1070, 2010.
21. S. D. Cunningham, W. R. Berti, and J. W. W. Huang, "Phytoremediation of contaminated soils,"Trends in Biotechnology, vol. 13, no. 9, pp. 393–397, 1995.
22. W. W. R. Berti and S. D. Cunningham, "Phytostabilization of metals," in Phytoremediation of Toxic Metals—Using Plants to Clean up the Environment, I. Raskin and B. D. Ensley, Eds., pp. 71–88, John Wiley & Sons, New York, NY, USA, 2000.
23. B. R. Sabey, R. L. Pendleton, and B. L. Webb, "Effect of municipal sewage sludge application on growth of two reclamation shrub species in copper mine spoils," Journal of Environmental Quality, vol. 19, no. 3, pp. 580–586, 1990.
24. S. L. Brown, C. L. Henry, R. Chaney, H. Compton, and P. S. Devolder, "Using municipal biosolids in combination with other residuals to restore metal-contaminated mining areas," Plant and Soil, vol. 249, no. 1, pp. 203–215, 2003.
25. C. Verdugo, P. Sánchez, C. Santibáñez et al., "Efficacy of lime, biosolids, and mycorrhiza for the phytostabilization of sulfidic copper tailings in chile: a greenhouse experiment," International Journal of Phytoremediation, vol. 13, no. 2, pp. 107–125, 2011.
26. S. H. Schoenholtz, H. van Miegroet, and J. A. Burger, "A review of chemical and physical properties as indicators of forest soil quality: challenges and opportunities," Forest Ecology and Management, vol. 138, no. 1–3, pp. 335–356, 2000.
27. P. H. Müller da Silva, F. Poggiani, and J. P. Laclau, "Applying sewadge sludge to Eucalyptus grandis plantations: effects on biomass production and nutrient cycling through litterfall,"Applied

and Environmental Soil Science, vol. 2011, Article ID 710614, 11 pages, 2011.

28. U.S. EPA, "The use of soil amendments for remediation, revitalization, and reusue," EPA 542-R-07-013, Office of Superfund Remediation and Technology Innovation (OSRTI), EPA/National Service Center for Environmental Publications, Cincinnati, Ohio, USA, 2007.
29. P. Janoš, J. Vávrová, L. Herzogová, and V. Pila ová, "Effects of inorganic and organic amendments on the mobility (leachability) of heavy metals in contaminated soil: a sequential extraction study," Geoderma, vol. 159, no. 3-4, pp. 335–341, 2010.
30. A. Lopareva-Pohu, B. Pourrut, C. Waterlot et al., "Assessment of fly ash-aided phytostabilisation of highly contaminated soils after an 8-year field trial—part 1. Influence on soil parameters and metal extractability," Science of the Total Environment, vol. 409, no. 3, pp. 647–654, 2011.
31. L. A. Junqueira-Teixeira, R. S. Berton, A. R. Coscione, and L. A. Saes, "Biosolids application on banana production: soil chemical properties and plant nutrition," Applied and Environmental Soil Science, vol. 2011, Article ID 238185, 8 pages, 2011.
32. D. Nash, C. Butler, J. Cody, et al., "Effects of biosolids application on pasture and grape vines in south-eastern Australia," Applied and Environmental Soil Science, vol. 2011, Article ID 342916, 11 pages, 2011.
33. C. Santibáñez, C. Verdugo, and R. Ginocchio, "Phytostabilization of copper mine tailings with biosolids: implications for metal uptake and productivity of Lolium perenne," Science of the Total Environment, vol. 395, no. 1, pp. 1–10, 2008.
34. W. E. Sopper, Municipal Sludge Use for Land Reclamation, Lewis Publishers, Ann Arbor, Mich, USA, 1993.
35. K. C. Haering, W. L. Daniels, and S. E. Feagly, "Reclaiming mined lands with biosolids, manures and papermill sludges," in Reclamation of Drastically Disturbed Lands, R. Barnhisel, Ed., pp. 615–644, Soil Science Society of America, Madison, Wis, USA, 2000.
36. INDAP, Compendio de Información Ambiental, Socioeconómica y Silvoagropecuaria de la IV Región de Coquimbo: Información

Para Proyectos de Desarrollo Regional, Universidad de Chile, Centro de Agricultura Ambiente (AGRIMED), INDAP-Prodecop IV Región, La Serena, Chile, 2001.

37. K. Lombard, M. O›Neill, A. Ulery, et al., "Fly ash and composted biosolids as a source of Fe for hybrid poplar: a greenhouse study," Applied and Environmental Soil Science, vol. 2011, Article ID 475185, 11 pages, 2011.
38. CIREN, Estudio Agrológico IV Región; Descripciones de Suelos, Materiales y Símbolos, Centro de Información de Recursos Naturales, Santiago, Chile, 2005.
39. J. F. Casale, R. Ginocchio, and P. León-Lobos, Fitoestabilización de Depósitos de Relaves en Chile. Guía 4: Marco Ambiental y Relaves Mineros Abandonados, Documento Técnico, Centro de Investigación Minera y Metalúrgica e Instituto de Investigaciones Agropecuarias, Santiago, Chile, 2011.
40. USDA, Survey Laboratory: Methods Manual, Investigations Report No. 42, Version 4.0, United States Department of Agriculture, Washington, DC, USA, 2004.
41. K. Gold, P. León-Lobos, and M. Way, Manual de Colecta de Semillas de Plantas Silvestres para Conservación y Restauración Ecológica, Instituto de Investigaciones Agropecuarias, Centro Regional de Investigación Intihuasi, La Serena, Chile, 2004, Boletín INIA no. 110.
42. P. M. Bleeker, A. G. L. Assunção, P. M. Teiga, T. De Koe, and J. A. C. Verkleij, "Revegetation of the acidic, as contaminated Jales mine spoil tips using a combination of spoil amendments and tolerant grasses," The Science of the Total Environment, vol. 300, no. 1–3, pp. 1–13, 2002.
43. A. D. Bradshaw and M. J. Chadwick, the Restoration of Land. The Ecology and Reclamation of Derelict and Degraded Land, University of California Press, Berkeley, Calif, USA, 1980.
44. M. Mench, S. Bussière, J. Boisson et al., "Progress in remediation and revegetation of the barren Jales gold mine spoil after in situ treatments," Plant and Soil, vol. 249, no. 1, pp. 187–202, 2003.
45. U.S. Environmental Protection Agency, Test Methods for Evaluating Solid Waste—Physical/Chemical Methods, Method SW- 486, Department of Commerce, National Technical Information Service, Springfield, Va, USA, 1997.

46. A. Sadzawka, M. Carrasco, R. Grez, M. Mora, H. Flores, and A. Neaman, Métodos de Análisis Recomendados para los Suelos de Chile, Instituto de Investigaciones Agropecuárias, Santiago, Chile, 2006.
47. R.Vulkan, F.J. Zhao,V. Barbosa-Jefferson et al., "Copper speciation and impacts on bacterial biosensors in the pore water of copper-contaminated soils," Environmental Science and Technology, vol. 34, no. 24, pp. 5115–5121, 2000.
48. U.S. Environmental Protection Agency, "Methods for chemical analysis of water and wastes,"EPA 60014-79-020, Washington, DC, USA, 1983.
49. J. P. E. Anderson, "Soil respiration," in Methods of Soil Analysis, Part 2. Chemical and Microbiological Properties, A. L. Page, Ed., pp. 837–871, Soil Science Society of America, Madison, Wisconsin, USA, 1982.
50. E. G. Gregorich, G. Wen, R. P. Voroney, and R. G. Kachanoski, "Calibration of a rapid direct chloroform extraction method for measuring soil microbial biomass C," Soil Biology and Biochemistry, vol. 22, no. 7, pp. 1009–1011, 1990.
51. G. P. Sparling, C. W. Feltham, J. Reynolds, A. W. West, and P. Singleton, "Estimation of soil microbial C by a fumigation-extraction method: use on soils of high organic matter content, and a reassessment of the KEC-factor," Soil Biology and Biochemistry, vol. 22, no. 3, pp. 301–307, 1990.
52. T. H. Anderson and K. H. Domsch, "Application of eco-physiological quotients (qCO_2 and qD) on microbial biomasses from soils of different cropping histories," Soil Biology and Biochemistry, vol. 22, no. 2, pp. 251–255, 1990.
53. J. A. D. Rienzo, F. Casanoves, M. G. Balzarini, L. González, M. Tablada, and C. W. Robledo,InfoStat Versión 2009, Grupo InfoStat, FCA, Universidad Nacional de Córdoba, Argentina, 2009.
54. W. G. Hopkins, "A scale of magnitudes for effect statistics," in A New View of Statistics, Internet Society of Sport Science, 2002, http://www.sportsci.org/resource/stats/effectmag.html.
55. M. E. Sumner, "Sodic soils: new perspectives," in Australian Sodic Soils: Distribution, Properties and Management, R. Naidu, M. E.

Sumner, and P. Rengasamy, Eds., pp. 1–34, CSIRO, Melbourne, Australia, 1995.

56. S. Brown, A. Svendsen, and C. Henry, "Restoration of high zinc and lead tailings with municipal biosolids and lime: a field study," Journal of Environmental Quality, vol. 38, no. 6, pp. 2189–2197, 2009.

57. D. N. Rietz and R. J. Haynes, "Effects of irrigation-induced salinity and sodicity on soil microbial activity," Soil Biology and Biochemistry, vol. 35, no. 6, pp. 845–854, 2003.

58. M. Sardinha, T. Müller, H. Schmeisky, and R. G. Joergensen, "Microbial performance in soils along a salinity gradient under acidic conditions," Applied Soil Ecology, vol. 23, no. 3, pp. 237–244, 2003.

59. G. Southam and T. J. Beveridge, "Enumeration of Thiobacilli within pH-neutral and acidic mine tailings and their role in the development of secondary mineral soil," Applied and Environmental Microbiology, vol. 58, no. 6, pp. 1904–1912, 1992.

60. C. Garcia, T. Hernandez, C. Costa, and M. Ayuso, "Evaluation of the maturity of municipal waste compost using simple chemical parameters," Communications in Soil Science and Plant Analysis, vol. 23, no. 13-14, pp. 1501–1512, 1992.

61. C. A. Campbell and R. P. Zentner, "Soil organic matter as influenced by crop rotations and fertilization," Soil Science Society of America Journal, vol. 57, no. 4, pp. 1034–1040, 1993.

62. J. A. Pascual, C. Garcia, T. Hernandez, J. L. Moreno, and M. Ros, "Soil microbial activity as a biomarker of degradation and remediation processes," Soil Biology and Biochemistry, vol. 32, no. 13, pp. 1877–1883, 2000.

63. C. Graciano, J. F. Goya, M. Arturi, C. Pérez, and J. L. Frangi, "Fertilization in a fourth rotationEucalyptus grandis plantation with minimal management," Journal of Sustainable Forestry, vol. 26, no. 2, pp. 155–169, 2008.

64. S. Brown, P. de Volder, H. Compton, and C. Henry, "Effect of amendment C : N ratio on plant richness, cover and metal content for acidic Pb and Zn mine tailings in Leadville, Colorado,"Environmental Pollution, vol. 149, no. 2, pp. 165–172, 2007.

65. Ch. Henry, D. Sullivan, R. Rynk, K. Dorsey, and C. Cogger, Managing Nitrogen from Biosolids, Washington State Department of Ecology, Northwest Biosolids Management Association, Wash, USA, 1999.

66. D. G. Brockway, D. H. Urie, P. V. Nguyen, and J. B. Hart, "Wastewater and sludge nutrient utilization in forest ecosystems," in The Forest Alternative for Treatment and Utilization of Municipal and Industrial Wastes, D. W. Cole, C. L. Henry, and W. L. Nutter, Eds., pp. 221–245, University of Washington Press, Seattle, Wash, USA, 1986.

67. P. Faúndez, Actividad Microbiológica Global en Suelos Acondicionados con Biosólidos Cloacales Frescos y Compostados con Residuos Forestales, Universidad de Chile, Facultad de Ciencias Agronómicas, Santiago, Chile, 2005.

68. T. Gea, A. Artola, X. Sort, and A. Sánchez, "Composting of residuals produced in the catalan wine industry," Compost Science and Utilization, vol. 13, no. 3, pp. 168–174, 2005.

69. D. Messerer, Sustratos Alternativos en la Propagación de Palto (Persea americana), Universidad Católica de Valparaíso, Facultad de Agronomía, Valparaíso, Chile, 1998.

70. C. Santibáñez, R. Ginocchio, and M. Teresa Varnero, "Evaluation of nitrate leaching from mine tailings amended with biosolids under Mediterranean type climate conditions," Soil Biology and Biochemistry, vol. 39, no. 6, pp. 1333–1340, 2007.

71. A. Navarro and F. Martínez, "Effects of sewage sludge application on heavy metal leaching from mine tailings impoundments," Bioresource Technology, vol. 99, no. 16, pp. 7521–7530, 2008.

72. I. Lamy, S. Bourgeois, and A. Bermond, "Soil cadmium mobility as a consequence of sewage sludge disposal," Journal of Environmental Quality, vol. 22, no. 4, pp. 731–737, 1993.

73. M. B. McBride, S. Sauvé, and W. Hendershot, "Solubility control of Cu, Zn, Cd and Pb in contaminated soils," European Journal of Soil Science, vol. 48, no. 2, pp. 337–346, 1997.

74. V. Antoniadis and B. J. Alloway, "The role of dissolved organic carbon in the mobility of Cd, Ni and Zn in sewage sludge-amended soils," Environmental Pollution, vol. 117, no. 3, pp. 515–521, 2002.

75. A. S. Coninck and A. Karam, "Impact of organic amendments on aerial biomass production, and phytoavailability and fractionation of copper in a slightly alkaline copper mine tailing,"International Journal of Mining, Reclamation and Environment, vol. 22, no. 4, pp. 247–264, 2008.

76. S. Klitzke and F. Lang, "Hydrophobicity of soil colloids and heavy metal mobilization: effects of drying," Journal of Environmental Quality, vol. 36, no. 4, pp. 1187–1193, 2007.

77. J. O. Miller, A. D. Karathanasis, and C. J. Matocha, "In situ generated colloid transport of Cu and Zn in reclaimed mine soil profiles associated with biosolids application," Applied and Environmental Soil Science, vol. 2011, Article ID 762173, 9 pages, 2011.

78. M. A. Al-Wabel, D. M. Heil, D. G. Westfall, and K. A. Barbarick, "Solution chemistry influence on metal mobility in biosolids-amended soils," Journal of Environmental Quality, vol. 31, no. 4, pp. 1157–1165, 2002.

79. F. Cabrera, R. López, A. Martinez-Bordiú, E. De Dupuy Lome, and J. M. Murillo, "Land treatment of olive oil mill wastewater," International Biodeterioration and Biodegradation, vol. 38, no. 3-4, pp. 215–225, 1996.

80. F. A. El-Gohary, M. I. Badawy, M. A. El-Khateeb, and A. S. El-Kalliny, "Integrated treatment of olive mill wastewater (OMW) by the combination of Fenton's reaction and anaerobic treatment,"Journal of Hazardous Materials, vol. 162, no. 2-3, pp. 1536–1541, 2009.

81. M. Stoller, "On the effect of flocculation as pretreatment process and particle size distribution for membrane fouling reduction," Desalination, vol. 240, no. 1–3, pp. 209–217, 2009.

82. C. I. Piperidou, C. I. Chaidou, C. D. Stalikas, K. Soulti, G. A. Pilidis, and C. Balis, "Bioremediation of olive oil mill wastewater: chemical alterations induced by Azotobacter vineladii," Journal of Agricultural and Food Chemistry, vol. 48, no. 5, pp. 1941–1948, 2000.

83. A. Nastri, N. A. Ramieri, R. Abdayem, R. Piccaglia, C. Marzadori, and C. Ciavatta, "Olive pulp and its effluents suitability for soil amendment," Journal of Hazardous Materials, vol. 138, no. 2, pp. 211–217, 2006.

84. A. A. Zorpas and V. J. Inglezakis, "Integrated applied methodology for the treatment of heavy polluted waste waters from olive oil industries," Applied and Environmental Soil Science, vol. 2011, Article ID 537814, 14 pages, 2011.
85. S. A. Pavan-Fernandes, W. Bettiol, and C. Clementi-Cerri, "Effect of sewage sludge on microbial biomass, basal respiration, metabolic quotient and soil enzymatic activity," Applied Soil Ecology, vol. 30, no. 1, pp. 65–77, 2005.
86. M. Lytle, F. W. Lytle, N. Yang et al., "Reduction of Cr (VI) to Cr (III) by wetland plants: potential for in situ heavy metal detoxification," Environmental Science and Technology, vol. 32, no. 20, pp. 3087–3093, 1998.
87. A. J. M. Baker and P. L. Walker, "Physiological responses of plants to heavy metals and the quantification of tolerance and toxicity," Chemical Specialization and Bioavailability, vol. 1, pp. 7–17, 1989.
88. D. C. Adriano, Trace Elements in Terrestrial Environments. Biogeochemistry, Bioavailability, and Risk of Metals, Springer, New York, NY, USA, 2001.
89. A. Kabata-Pendias and H. Pendias, Trace Element in Soils and Plants, CRC Press LLC, Boca Raton, Fla, USA, 2001.
90. D. J. Walker, R. Clemente, A. Roig, and M. P. Bernal, "The effects of soil amendments on heavy metal bioavailability in two contaminated Mediterranean soils," Environmental Pollution, vol. 122, no. 2, pp. 303–312, 2003.
91. R. Clemente, Á. Escolar, and M. P. Bernal, "Heavy metals fractionation and organic matter mineralisation in contaminated calcareous soil amended with organic materials," Bioresource Technology, vol. 97, no. 15, pp. 1894–1901, 2006.
92. E. J. M. Temminghoff, S. E. A. T. M. van der Zee, and F. A. M. de Haan, "Copper mobility in a copper-contaminated sandy soil as affected by pH and solid and dissolved organic matter,"Environmental Science and Technology, vol. 31, no. 4, pp. 1109–1115, 1997
93. National Research Council, Nutrient Requirements of Domestic Animals, National Academy of Science, National Academy Press, Washington, DC, USA, 7th edition, 1984.

94. I. González, V. Muena, M. Cisternas, and A. Neaman, "Acumulación de cobre en una comunidad vegetal afectada por contaminación minera en el valle de Puchuncaví, Chile central," Revista Chilena de Historia Natural, vol. 81, no. 2, pp. 279–291, 2008.

95. A. Anic, L. F. Hinojosa, J. Díaz-Forester et al., "Influence of soil chemical variables and altitude on the distribution of high-alpine plants: the case of the Andes of central Chile," Arctic, Antarctic, and Alpine Research, vol. 42, no. 2, pp. 152–163, 2010.

96. A. Fahn and D. F. Cutler, Xerophytes, Gebruder Borntraeger, Berlin, Germany, 1992.

Chapter 2

Property Rights and Sustainable Irrigation: A Developing Country Perspective

Ruth Meinzen-Dick

International Food Policy Research Institute, 2033 K St NW, Washington, DC 20006, USA

ABSTRACT

While the role of secure property rights contributing to sustainable natural resource management is increasingly recognized, translating that into practice is more challenging, especially in developing countries. This article presents a framework for understanding the role of property rights for effective irrigation systems and then explores the complexity of property rights to land, water, and infrastructure and their underlying institutions. Understanding property rights in practice requires acknowledging legal pluralism—the coexistence of many types and sources of law, which can be used as the basis for claiming rights over the resources. Property rights do not necessarily imply full ownership, but are composed of different bundles of rights that may be held by different claimants—the state, user groups, families, or individuals. These rights are critical for the authority, incentives, and resources for irrigation operation and maintenance. As resources become more scarce, property rights systems need to adapt to reduce conflict and provide incentives for saving water. However, efforts to improve irrigation by changing property rights systems have often failed because they have not recognized the difficulty of transplanting property rights systems from one place to another. Institutional change needs to be seen as an organic process, building on existing norms and practices, rather than as an exercise in social engineering.

INTRODUCTION

Experience with the past 30 years of irrigation has shown that technology alone is not sufficient to ensure productivity gains, let alone sustainability. In many cases, the technologies were not adopted or maintained, or the poor, women, and other marginalized groups were excluded from the benefits of technologies. Appropriate institutions are needed to accompany technologies for sustainable irrigation.

Among these key institutions, property rights play a particularly important role. When resources are abundant, there is little need to define property rights, but as the resource becomes more scarce, users of the resource compete and even come into conflict. In such contexts, there is pressure to define property rights over the resource, to clarify expectations, and assign both rights and duties (Otsuka and Place, 2001 and Young, in press).

While the contribution of secure property rights to sustainable natural resource management is increasingly recognized in calls for policy reforms, translating that into practice is more challenging (Deininger, 2003). This applies to almost all natural resources, including land, forests, and fisheries, but even more so to water, which is a mobile and often fugitive resource, difficult to measure or control, with many different uses and values, both economic and noneconomic (Rogers et al., 1995). Efforts to "improve" irrigation by changing property rights systems have often failed because they have not recognized the difficulties involved in institutional change, especially of property rights.[1]

While water rights, and especially water rights reforms, are complex in any context, the challenges are particularly severe in developing countries.[2] A major reason for this is the frequent lack of state capacity to carry out the reforms as envisaged (as described by Garduno, 2005 for Mexico or Schreiner, 2013 for South Africa). However, customary water rights are also likely to be particularly strong in developing countries, often varying from one context to another, and differing from state law (von Benda-Beckmann et al., 1998). Moreover, as North (1990) notes, institutional change is path dependent: it is inherently shaped by the history of a particular place. This means that transplanting property rights systems from one place to another will not yield the same results, as shown by Solanes (2013) in his analysis of efforts in the 1990s to pass a water law in Peru that was modeled after Chile's legal framework of highly privatized water resources.

Thus while property rights can play a crucial role in sustainable irrigation in developing countries, any reforms to achieve this aim need to be based on a sound understanding of property rights to land, water, and infrastructure and their underlying institutions. Of these, land rights have been dealt with most extensively, both in scholarship and in policy (see, for example, Deininger, 2003 or FAO, 2012). The focus of this article is therefore on rights to water and secondarily irrigation infrastructure. However, rights to all three of these key resources are inextricably intertwined (see Hodgson, 2004).

This article presents a framework for understanding the role of property rights for effective irrigation systems and then explores the complexity of these property rights, with particular reference to conditions in developing countries. Applying the framework helps to identify the situations in which institutional issues of coordination and property rights are likely to be important for sustainable irrigation (and where, conversely, they are less likely to require much attention). Subsequent sections of the article discuss some of the institutional complexities, particularly of property rights, and their implications for programs seeking to improve the sustainability of irrigation systems.

KEY INSTITUTIONS FOR SUSTAINABLE IRRIGATION

Institutions are the "rules of the game" that govern behavior in society (North, 1990:1). Fig. 1 illustrates the importance of two types of key institutions for irrigation and agricultural water management: those that provide coordination and property rights. The vertical axis illustrates the spatial scale of a technology, from an individual plot, through a whole farm, to one that covers several farms, a village, to a region. All approaches that are above the scale of the individual farm require some form of coordination—either by local organizations, the state, or the market. For example, a drip irrigation kit may be adopted by an individual small farmer, by his or her self. Even a well may serve just one farm, but where holdings are very small and tubewells have large capacity, farmers may join together to buy and operate a tubewell, or the state may install and operate it, or one farmer can install it and sell water to neighbors. As we move up the spatial scale, a small check dam may serve a group of farmers. A watershed management program may serve several communities. Canal irrigation systems and reservoirs can serve up to thousands of farmers, and even cross provincial lines. Finally, transboundary river basins cross national boundaries. In each case, some form of coordination is required to govern provision and expropriation: to ensure that the infrastructure is built and maintained, and to allocate and distribute the water among users of the same source, and settle disputes. The greater the spatial scale, the higher the level of coordination that is needed, as indicated by the corresponding arrow on the right side, to balance the spatial arrow on the left. That coordination may be provided by the state or by collective action (or, in some cases, even by the market).

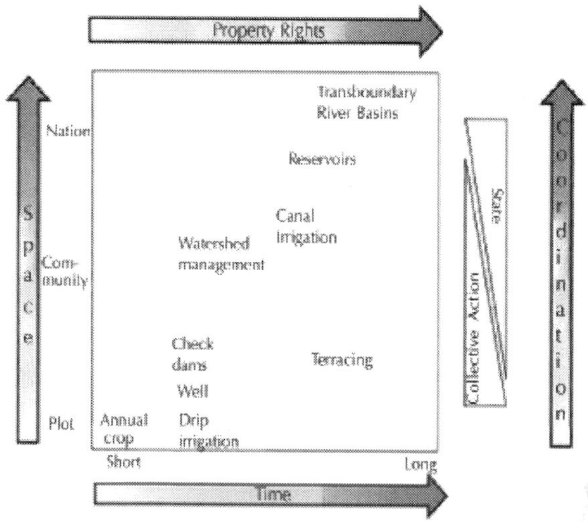

Figure. 1. The role of property rights and coordination institutions for sustainable irrigation.

Similarly, the lower horizontal axis indicates the permanence of a technology or approach, or the time frame between when an investment is made and its returns are realized. The longer the temporal scale, the greater the need for property rights to provide authorization and incentive to make the investment (Knox et al., 2002). Even a tenant or a wife without independent land rights can install a drip kit, but may not be allowed to install a treadle pump or tubewell, because she does not have the decision-making (management) rights over that land, and she may not have the incentive to install and maintain terracing or drainage systems for salinity control. The arrow on the upper horizontal axis indicates that secure property rights become more important as the time frame of irrigation practices increase. Even those with decision-making rights may not have the incentive to make long-term investments unless they have secure tenure, to know that they will benefit from the investment. Although security of tenure is often thought of in terms of rights to the land, people may not be willing to invest in irrigation systems if they do not also have secure rights to the water that will give them the returns. This has been the problem with many irrigation management transfer systems, where farmers were expected to bear the costs, without secure rights to the water from the systems.

While the exact location of any type of irrigation on Fig. 1 would depend on the size of the farms and the scale, as well as the cost/ return ratio, of the particular technology, this framework provides a useful starting point to ask which institutions are likely to be critical. As water uses increase, even seemingly "independent" water users will be affected by, and have an impact on, other water uses and users. For example, Asquith (2006) describes how immigration and more intensive land use upstream had impacts downstream in the Los Negros Valley of Bolivia. This increasing interaction between uses within basins calls for better institutional arrangements to coordinate water uses (as described by Asquith, 2006). Property rights are important in this regard because they clarify who can use and manage the land, water, or infrastructure, and what responsibilities they have toward the resource and toward others. Coordination institutions help to set and enforce those rules about the allocation of water and responsibilities. These issues are most apparent in surface water flows and irrigation, but also applies to groundwater irrigation. The following sections provide more details on the institutions for coordination and property rights, with special attention to their implications for sustainable irrigation.

COORDINATION INSTITUTIONS

The example of the tubewell cited above illustrates that coordination functions can be provided by the state (a public tubewell that supplies many farms), collective action (farmer group) or markets (farmer selling, buying or trading water). How well each of those institutions functions will determine whether farmers receive adequate and timely water supplies. For example, in the 1970s and 1980s there were numerous state-run tubewells in India, Bangladesh, and Pakistan that should have allowed economies of scale by using large-scale pumps to irrigate many surrounding farms. These large pumps offer greater potential technical efficiency over small-scale pumps. But these potential technical efficiency gains were often not realized because of poor operational performance of the government agencies running and maintaining the pumps. Instead of government-operated tubewells, a group of smallholder farmers may get together to make the investment in drilling the well, installing a pump, and operating it (collective action coordination), or an individual farmer may install and operate the well, and sell water to neighbors. Shah and Raju (1988) argue that groundwater markets are much more effective than state tubewells in serving farmers' water needs in India. Meinzen-Dick (1996) compared the performance of shared tubewells and groundwater markets in Pakistan, finding that smallholder farmers were quite well served by groundwater markets, but when the water became very scarce, the well owners would deny water to water purchasers, who had little recourse against more powerful well owners.

Which institution is most appropriate depends on the particular conditions—e.g. scale, technical sophistication of the technology and the farmers, cultural factors (social capital, market orientation), and state capacity. In general, the advantages of the state are greatest at the largest scale, and collective action at more localized levels (although there are exceptions). Markets are highly variable in whether they provide effective coordination among smallholders, but seem to come in where machinery is involved, where it is relatively easy to measure water deliveries and exclude those who do not pay (as from a tubewell or piped system), and where users have alternative sources so that there is competition and they are not dependent on one provider.

If group-based (collective action) approaches are selected by programs for water management or technology dissemination, it is important to look beyond the formal rules and membership roles to see whether the group is actually acting collectively: too many water user organizations exist only "on paper" and do not actually work together. Moreover, it is important to consider who is included and excluded from active membership and decision-making. This means asking about women as well as men, land owners and tenants, farmers and other water users (e.g. fishers, livestock keepers, home gardens, domestic users, other enterprises). There may be formal as well as informal barriers to participation, different motivations and returns to be considered. There are indications that organizations with active participation of men and women may be more effective than single-sex organizations in managing resources like water because they draw upon the skills and resources of both, but establishing active mixed organizations is also more difficult than for single-sex organizations, especially where there is high gender segregation in the society (Pandolfelli et al., 2008). All of these factors should be considered in relation to the objectives of the program when identifying the appropriate strategy for establishing or working with groups for irrigation. Unless there is a strong basis of collective action, group-based approaches may require significant investment of time and facilitation to build up the necessary social capital to be effective. Unless the groups are relatively homogeneous with active participation of men and women, further investment may be required to get diverse groups to work together. Where poor literacy and numeracy exist, adult education programs may be needed to support group functioning as well as increase overall human capital for poverty reduction.

Many irrigation management transfer programs that have sought to use farmer organizations for water management (Garces-Restrepo et al., 2007). One clear lesson from this and other irrigation management reforms is that there are no universal formulas or blueprint approaches; rather there is a need for adaptation to local conditions, and as they change over time (Merrey et al., 2007). The triangles on the right-hand side ofFig. 1 are not only an indicator of the broad comparative advantages of state and collective action institutions at each level, but also as a reminder of the overlap of both types of institutions, and the need for them to work together, as in co-management approaches. Instead of the state crowding out collective action, or local groups having to develop all technical capacity on their own, it is preferable to seek collaboration. For example, state agencies may provide technical backstopping, but recognize that farmers have important knowledge of local conditions, or state agencies may facilitate access to external resources or help resolve disputes. This is the premise behind a growing number of catchment forums (as in South Africa) and other multi stake holder platforms. But even this is not straightforward: as Komakech and van der Zaag (2013)indicate in their study of catchment forums in Tanzania that a detailed study of local institutional arrangements is needed as a starting point, with attention to power structures, including whether large farmers, hydropower, or other influential actors will (or will not) participate.

Similarly, there are options for combining market mechanism with user groups and the state. In identifying the potential for water markets, Rosegrant and Binswanger (1994) recognize the importance of user groups to provide coordination among smallholders involved in water trades above the very local level, and overall state regulation. Similar combinations of market, collective action, and state institutions are also seen in payment for environmental services or watershed management programs that compensate upstream users for changing their water use patterns to improve the flow, quality, or timing of water downstream (e.g.Asquith, 2006, Swallow et al., 2002 and Swallow et al., 2006). Because the coordinating institution will need to have at least some control over the infrastructure and even the water itself, the choice of coordinating institution (or institutions) also has strong implications for property rights. The implications of this are discussed in more detail in the following section on property rights.

UNDERSTANDING PROPERTY RIGHTS

There are numerous definitions of property rights, including: "the capacity to call upon the collective to stand behind one's claim to a benefit stream" (Bromley, 1991:15), "the claims, entitlements and related obligations among people regarding the use and disposition of a scarce resource" (Furubotn and Pejovich, 1972:1139–1140), or "the social institutions that define or delimit the range of privileges granted to individuals to specific assets, such as parcels of land or water..." (Libecap, 1998:77).

Key elements in these definitions are that property rights are social institutions: relationships among people with regard to resources. They entail recognized claims and obligations, rights and duties or responsibilities (Commons, 1968). To be valid, rights must be accompanied by a duty of others to respect those rights. The second set of duties or responsibilities are the responsibilities of the right-holder, which may be related to*use* or *provision*: to use the resource in certain ways (e.g. without polluting the water), or to provide for the construction or maintenance of facilities (e.g. irrigation infrastructure).

Ribot and Peluso (2003) distinguish between *access* to resources (e.g. being physically able to get water) and *rights* (entitlements). A tail end farmer may have a right to water, but might not be able to get access because those upstream take too much. A farmer may also have access due to physical location, force, or even stealth, even if he or she does not have a right to the water. Many water users do not have formalized rights to the water they depend on for their livelihoods. Pradhan and Pradhan (2000) provide an example of this distinction between access and rights from the hills of Nepal, in which irrigators from one system would allow farmers from another area to use water under what may be referred to as "tolerated use", as long as the farmers from the second system did not claim rights over that water. However, this access was not enough for the latter farmers, who used a range of strategies to claim rights over that water. Pradhan and Pradhan (2000:201) find that this has broader applicability to many farmer-managed systems:

> "Wherever possible they divert water from new (uncontested) sources, and if this is not possible they try to acquire water from sources used by others by negotiating, disputing, 'stealing' water on the sly, or forcefully acquiring water, often using political or administrative connections. The farmers are not satisfied with just acquiring water; their long-term goal is to legitimize their access to the water source, that is, to establish rights to the water... to get their claims to a share of the water from a water source accepted by other users and competitors."

With regard to irrigation systems, there are three types of property rights that are important: to the land, the infrastructure, and the water. All three of these resources are necessary for productive use of the water, and how rights over these resources are distributed will affect both the amount and distribution of benefits.

Recognizing property rights to water (as well as irrigation infrastructure) often requires going beyond simplistic, legalistic, or externally-imposed assumptions. When discussing water rights with government officials, I have often heard "There are no water rights here" or "The State owns all water." Both of these statements derive from a very narrow understanding of property rights. Conventional images of property rights often derive from images of unencumbered land rights, with fixed boundaries, within which the individual "owner", backed by a state-issued title, can do whatever he (or more rarely, she) chooses with the resource.[3] Those boundaries are seen as rigid and unchanging, dividing people between those on the inside and outside. Certainly these conditions rarely apply to water, particularly in developing countries.

But this view of property rights is flawed, even for land. As noted above, property rights are above all, social institutions, which define relationships among people. They connect more than they divide. Rights are never unencumbered. Instead of "ownership" that implies rights to do whatever one wants to the resource, it is better to think of "bundles of rights" that may be held by different claimants, including individuals, households, water user groups, communities, and the state. Indeed, the inter linkages between these different claimants on the resource can even be conceptualized as a "web of interests" (Arnold, 2002). Furthermore, all rights do not derive from the state: there may be many different sources of property rights. Each of these aspects apply, to some extent, even to immobile property in industrialized countries. But they are especially relevant to property rights in developing countries, and especially to rights over a fluid and often fugitive resource such as water. The following sections provide more detail on legal pluralism and bundles of rights, as they pertain to irrigation systems.

Legal Pluralism

Understanding property rights in practice requires acknowledging legal pluralism—the coexistence of many types and sources of law, which can be used as the basis for claiming rights over the resources or regulations on what can be done with them.[4] This is often thought of in terms of "statutory" and "customary" law, but it is more complex than that. State law itself may be plural, with different – and even inconsistent –.laws issued for different types of water use. Komakech et al. (2012) describe how water use permits are interpreted and enforced differently by various administrative units in Tanzania. Customary law is yet more diverse, differing between communities (however defined) and, because it is often unwritten, even within a community. Moreover, although "custom" is often interpreted as being rigid and unchanging, in fact "customary" law changes, often quite rapidly, to the extent that they may be referred to as "living law" instead (Ehrlich, 1936).

Beyond this conventional dichotomy there are international laws and treaties, such as the Ramsar Convention governing wetlands,[5] or international treaties governing shared watercourses. Each irrigation project generates its own regulations, such as which parts of the command area are authorized to receive water in each season, from which date, and what the water users are required to do (and are prohibited from doing). Water users' associations generate their own rules regarding the manner in which different people may draw water, and what each needs to provide for the system. Because of the fundamental importance of water for life itself, religious laws and rules also create the basis for water rights, as illustrated in Islamic water law

(Faruqui et al., 2001) or the Balinese *subak* irrigation systems, which derive much of the authority for their detailed system of water rights from a series of temples, from the local field canal up to the whole watershed (Lansing, 1991). Even local norms and sayings can become the basis for claiming water rights. For example, in Kenya Onyango et al. (2007) relate how a local proverb that "even the hyena has a right to water" is the articulation of a norm that no one should be denied water, therefore anyone should be allowed to cross even private property to access water.

These different bases for property rights do not exist in isolation. Rather, the various forms of laws or rules overlap, interact, and influence each other. A change in international law may change state law to bring the latter into compliance; a change in national law may end up affecting custom; but also, customary or religious rules may shape statutory law.

However, property rights are only as strong as the underlying institutions – whether state, community, religious, or other – that shape, endorse and enforce the rights. Rights that are not widely known, recognized, or enforced, have little meaning. Meinzen-Dick and Nkonya (2007) describe these overlapping frameworks as "force fields" of varying strength. In the capital city, among a highly educated population that can call upon the courts or other government officials to enforce it, state law is generally stronger than in the hinterland, where even the officials – let alone the general public – are often unaware of the provisions in statutory law. Each type of property rights has associated bearers of the law, such as state authorities, elders, religious leaders, or irrigation system officials. The strength of their authority certainly affects the strength of the property rights that they enforce. However, none of these have the ability to be everywhere and govern every water use or every act to maintain (or damage) an irrigation system. Property rights and obligations also have to be known by those immediately involved, which requires internalizing the rules.

This has important implications for the sustainability of irrigation systems. For any type of property rights to provide security and incentives for careful management of the water or infrastructure, the rules need to be clearly understood as well as enforced. Four of Ostrom's (1990) well-known "design principles" for the governance of irrigation systems relate to this: clearly defined boundaries, monitoring, graduated sanctions, and conflict resolution mechanisms. These principles were developed from a review of many studies of irrigation systems (as well as other types of common property). Cox et al. (2010) validated these principles based on many more such studies and found that it is important not only that the resource and its use are monitored, but that the monitoring be done either by the users themselves or by someone who is accountable to the users.

Bundles of Rights

As noted above, property rights to any resource almost never convey total rights, as is conjured up in notions of "freehold" or "ownership." Rather than holding that up as a standard, it is better to look at the different particular rights that different stakeholders have, which vary enormously, not only between different societies, but also between users within a particular locale. For comparative purposes, these are often identified as *bundles of rights,* but even how these bundles are defined varies. Schlager and Ostrom (1992)define these bundles in terms of *use* rights of access and withdrawal, and *decision-making* (or control) rights of management, exclusion, and alienation. Others, following older Roman tradition, group the bundles as:*usus* rights to *use* the resource; *usufructus* rights to derive income from a resource; *abusus* rights to change the resource; and *alienation* rights to transfer the rights to others (see Eggertsson, 1990, Alchian and Demsetz, 1972 and Scott, 2008). Combining the two classification systems, the following provide examples of the bundles of rights that are particularly relevant for sustainable irrigation.

- *Access* refers to rights for non consumptive use. The "right to ramble" or cross someone's land for a walk is a classic example of such access. Swimming in a reservoir would be another such example. Although access rights should be non consumptive, there may be crowding effects or wearing down of infrastructure if too many people access land or water resources.
- *Withdrawal* refers to rights to take resource units, such as consumptive uses of water or taking produce of the land.
- *Usufructus* rights to earn income from the resource are not included in the Schlager and Ostrom (1992)framework, but can be important where the state charges a water licensing fee. [6]

- *Management* covers rights to make rules and modify the resource, notably setting water delivery patterns or conducting maintenance.
- *Exclusion* refers to rights to determine who can use the resource, and the transfer of those rights. In the case of irrigation, the right to decide who can draw water for irrigation, and whether other users can also access the water, infrastructure, or adjacent land are all crucial exclusion rights.
- *Alienation* refers to the rights to transfer management and exclusion rights to others, such as by sale or lease.

"Ownership" or "complete property rights" are often interpreted as holding all of these bundles of rights. If the term "water rights" conjures up these images, then it is quite understandable that the concept would not seem to apply to water, especially in developing countries.[7] This is especially true if one looks only for formal rights issued by the state. Recognizing this, Hodgson (2013) argues for a focus on *tenure* rather than*rights*:

"A key advantage of couching the issue of water allocation in terms of tenure, rather than simply in terms of water rights, is that it facilitates recognition of the fact that a complex range of different, yet inter-linked, types of relationship exist between people with respect to water resources. More specifically, the question of water tenure goes beyond water rights and water law (Hodgson, 2013:1)."

This is a useful reminder that the focus needs to be on the institutional arrangements, rather than on specific formulations of what is recognized as property rights.

DISCUSSION: PROPERTY RIGHTS AND IRRIGATION MANAGEMENT

Attention to property rights in developing country irrigation systems came to prominence in the 1980s and 1990s in response to concerns with irrigation system maintenance and water scarcity. This section analyzes the policy debates and trends around each of these areas, and the factors that influence security of tenure.

Maintenance

Many irrigation systems have been developed by government agencies, but the performance of these systems has often been far below expectations. With the fiscal crisis of the state in the 1980s, funding for system maintenance was often scarce, causing further deterioration of systems. At the same time, studies of farmer-managed irrigation systems showed that in many cases these performed better than state-managed systems, because the farmers themselves undertook maintenance and irrigation staff were responsible to the farmers (Lam, 1998, Ostrom, 1992, Tang, 1992 and Uphoff, 1986).

Irrigation rehabilitation projects have identified one of the causes of these problems with government-managed irrigation system as being the tendency of the irrigators to feel that the responsibility for maintenance of the systems rested with the government, but the government agencies lacked the resources or capacity to do adequate maintenance. In order to deal with this, many government officials and donor-assisted irrigation programs have called for farmers to develop a "sense of ownership" over the irrigation systems so that they will take care of them (e.g. Vermillion, 1987).[8]Coward (1986) explicitly recognized the link between property rights and management, but focused on the effect of creating and maintaining property together as creating the "social glue" that would strengthen irrigator groups to maintain their irrigation.

Three reasons why property rights are important for maintaining irrigation systems can be summarized as authority, incentives, and resources (Meinzen-Dick, 2002).

- *Authority:* Unless farmers are recognized as having management rights over the infrastructure and water, they do not have the authority to make any changes in the irrigation systems, including maintenance.
- *Incentives:* Property rights, especially *use* and *usufructus* (income) rights, can provide incentives to maintain irrigation systems because they create confidence that the holder of the rights will reap the future benefits of investment and careful maintenance, and bear the losses incurred by misuse of the resources.
- *Resources:* Control over water, land and infrastructure, including *usufructus* rights to earn income from them, can

generate resources (cash or labor) for maintaining irrigation systems.

In most government irrigation systems the state claims the rights of management, exclusion, and earning income, and irrigators only have recognized use rights, often upon payment of an irrigation fee that is then supposed to fund operation and maintenance. However, there is often no direct link between the fees paid by irrigators and the budgets available for the irrigation agency, so the resources are not made available to sustain the irrigation systems. At the same time, farmers lacked the authority to undertake maintenance, and often had little incentive to do so. Recognizing these problems, a number of irrigation management transfer programs have included ceremonies for transferring "ownership" of the infrastructure of surface irrigation systems from the government to farmers' organizations (Vermillion, 1997). However, without rights to the water as well, the infrastructure itself is of little value, and even is a burden to maintain. The water rights are key assets, because they determine both the level and distribution of benefits from irrigation.

This is not to imply that secure property rights will necessarily lead to adequate maintenance. If the holder of the rights (farmers or the state) has the authority but not the technical expertise, if the returns to irrigation are not high enough to provide incentives, or if the costs of maintenance exceed what the usufructus rights can generate, having the rights will not be sufficient. However, the role of property rights merits greater attention than it has received in most discussions of irrigation maintenance or programs to improve irrigation maintenance.

Water Scarcity

As noted above, as water becomes scarce, it brings users into greater competition with each other. Even relatively water-abundant systems may have scarcity at some times, whether during a dry season or drought year, or as water use increases. Clear property rights define allocation rules for who may use how much water, in what ways, at what time and place. By clarifying expectations, these rules can help reduce conflicts. But a further reason for attention to water rights in the context of scarcity is to provide incentives for efficient use and water conservation. Rosegrant and Binswanger (1994) argue that tradable property rights would create incentives for irrigators (and other water users) to conserve water, because what they did not use could be sold to other uses (including to the environment). Thus, rather than requiring some users to give up water, which often causes opposition, this would allow voluntary transfers from less efficient users to those who could use it more productively, with those who give up water receiving payment. It is beyond the scope of this paper to review empirical experiences with water markets,[9] but it is important to note that tradable water rights hinge upon vesting the right to transfer water to others in the water users, rather than in the state. While irrigation management reform programs in many countries have been willing (albeit sometimes reluctantly) to transfer to the users the use, usufruct, management, and exclusion rights, vesting alienation rights in the users has been even more controversial.

Some farmer-managed irrigation systems, as in Nepal and Bali, have developed water rights systems in which proportions of the flows are separable from the land itself. In these cases, farmers have the incentive to use the water efficiently to irrigate as much land as they can with their shares of water. Zekri and Al-Marshudi (2008) describe proportional water rights systems in centuries-old *falaj* systems in Oman that even reserve a share of the water rights to be auctioned to support the system operation and maintenance—thereby combining resources for sustaining the system and tradable water rights in a context of water scarcity.

The question of how water rights systems can adapt to growing scarcity, either through proportional allocations, tradable water rights, or other mechanisms, is likely to become increasingly important with increasing water use and climate change. Sharing ideas from one place to another can assist in adaptation, but copying systems of water rights is not likely to succeed, because it does not account for the variations in the physical and institutional environment. For example, Chilean water law, with tradable water rights, has been promoted as a model for other countries, but Bauer (2004) shows that the Water Code was shaped by the particular history and conditions of Chile, that are not found elsewhere, and even within that context, the Chilean water law has key shortcomings as well as strengths that need to be understood.

Security of Tenure

It is not only the content of the rights that matter, but the security with which they are held. Especially the strength of incentives to sustain irrigation systems depends on how confident right-holders can be that they will hold that right in the future. Key aspects of tenure security include the expected duration of the right and its robustness.

Duration of water rights are complex because of the inherent fluctuations of the water cycle; these are becoming more severe and uncertain with climate change. Thus, it is difficult to know, from one year to another, whether a water right can be realized the next year. Here infrastructure can increase water tenure security, by providing storage (reservoirs or groundwater) to buffer against short-term fluctuations.

Robustness of property rights refer to their ability to withstand challenges. Some of these challenges come from the physical environment (e.g. weather fluctuations or landslides that destroy infrastructure). Other challenges, however, come from the institutional environment. For example, immigrant communities may begin using water from the same source as other right-holders, with some basis for claiming that right. Or the state may decide to extend an irrigation system, creating new claimants, often to the detriment of existing users.

In some cases, robustness of property rights comes from being able to resist such challenges (e.g. by exercising exclusion rights to prevent new claimants from taking water away from the original right-holders). State recognition can play an important role in creating tenure security, especially where the state has enforcement capacity. Yet in many parts of Africa, in particular, even land is held in customary tenure without formal state recognition, and the costs of developing full cadastres and titling programs are very high, often prohibitively so because of the cost of demarcating property, registering it, settling disputes, and then keeping any registries up to date. It is even more expensive and difficult to develop registries of water rights, especially for millions of smallholder farmers, fishers, or other small-scale water users (van Koppenet al., 2007). Hodgson (2013) notes that many minor uses of water are exempt from requirements to register, especially for basic livelihood needs. But this can also create insecurity because it denies those users of state protection. This is particularly an issue in cases where so-called "land grabs" by outside investors have also become "water grabs", taking water away from existing users. For example, Bues and Theesfeld (2012)analyze how large-scale farms in Ethiopia took water away from farmer-managed irrigation systems, because of the outside investors' greater power and ties to the state.

Robustness also comes from being able to adapt to new situations, particularly changes in the hydrologic regime. Formalized state-issued rights may provide less adaptive capacity than systems in which farmers are able to play a greater role in setting and modifying the rules, i.e. when users have management rights.

Both infrastructure and institutions can play a role in increasing security of tenure, especially over water. The role of storage (infrastructure) in helping people respond to droughts has already been mentioned. Pipes that reduce transmission losses can increase confidence that users will receive the water they are entitled to. Meters and information technology (e.g. monitoring others' water uses) can increase users' confidence that if they follow the rules, others will as well because all will be monitored and can be caught if they break the rules. But these technologies do not work on their own. Someone has to operate the reservoirs, read the meters, or provide the information; hence institutional arrangements are needed to ensure the infrastructure delivers.

South Africa's experience with the National Water Act (RSA, 1998) is instructive of the challenges of developing an equitable water rights system that responds to water scarcity, giving due priority to environmental and basic domestic water uses. Despite active consultation in developing the law, implementation has proved a challenge, especially with limitations of state capacity to monitor and enforce (Schreiner, 2013), continuing resistance by powerful water users who stand to lose from the reforms, and limitations on the ability to redress power asymmetries through water alone, without also addressing land rights (Kemerink et al., 2011).

Rather than looking for any prescriptions for property rights reforms, it is more realistic to recognize that changes are likely to be a process of what Cleaver (2001:26) refers to as: "institutional bricolage", a process by which people consciously and unconsciously draw on existing social and cultural arrangements to shape institutions in response to changing situations." An important aspect of this understanding of institutional change—whether in property rights or coordination institutions, is that people themselves are actors, shaping the outcomes, with outsiders (including the state) playing variable roles.

CONCLUSIONS

The development of irrigation systems that serve more than one farm require coordination systems and some form of property rights to at least identify who has what rights to use, manage, and exclude others from the associated land, infrastructure, and water. Even for individual wells and rainwater harvesting, as more and more people start to irrigate, they impact other users in the watershed, and some form of coordination and rules governing water use can help prevent conflicts and give users some assurance that their investments in irrigation will be sustainable.

However, the state is not the only source of coordination and property rights; collective action through customary institutions, newly developed user groups, and even market mechanism can all play a role. Nor do property rights need to include complete "ownership": particular bundles of rights can be held by the state, user groups collectively, or different individual users. Use rights of access and withdrawal, as well as*usufructus* rights to earn income from the resources can provide incentives for investing in and maintaining irrigation systems; management and exclusion rights provide authority to do so; *usufructus* rights provide revenue for operation and maintenance. Alienation rights that are transferrable, either seasonally (rental) or in perpetuity (sale), can provide incentives for those with existing water rights to conserve their use because they can gain income from transferring the unused water to others.

What matters is the security of these rights—the confidence that they provide that others will fulfill their duty to respect those rights. Rights are only as strong as the institutions that stand behind them. Where the state has very little effective capacity – as in remote regions or when the state is weak – then customary rights are often stronger, especially if their legitimacy is reinforced by widely held norms with internalize self-enforcement so they do not have to be explicitly enforced by others. But customary rights alone may not be secure in contexts where the state decides to reallocate water (or land) to other users.

As resources become more scarce, property rights systems need to adapt. Irrigation development projects inherently change property rights, whether explicitly assigning water and land rights, or implicitly by generating rules about who can do what with the water and infrastructure, or simply by creating new "facts on the ground," of where and when water is available, which forces changes in property rights. Many irrigation management transfer projects have included formal ceremonies to confer on the farmers rights over the infrastructure (frequently only the onerous duties to maintain the infrastructure without real rights over its actual use and disposal), but rarely the water. The creation of transferable property rights, as in Chile or Australia, have been discussed but rarely implemented in developing countries (beyond informal markets for groundwater in South Asia).

Recognition of the need for coordinating institutions and property rights for sustainable and efficient management of irrigation is an important step. But identifying the important institutions is relatively easy compared to ensuring that these are in place. Efforts to improve irrigation by changing property rights systems have often failed because they have not recognized the difficulty of transplanting what is seen as clear property rights systems from one place to another. Rapid reforms can be counter-productive. They are unlikely to be fully implemented as planned, and create uncertainty and resentment in the process. Negotiation with stakeholders, looking for ways to compensate those who might lose out, leads to more legitimacy of the reforms.

Institutional change is possible, but it needs to be seen as an organic process, building on existing norms and practices, and recognizing that local people will be active in adapting – or even resisting – external interventions. Analogies of "social engineering" have been misplaced, because they imply a mechanistic approach. What is required is a more nuanced approach, which may involve adaptation to the physical and institutional environment. A good strategy is to start with people's experiences with water—how they access it, what claims they make for their different water uses. This will help to identify the relevant legal frameworks to address. Then an intervention can work to strengthen the claims of poor people for their important water uses. In many cases, water rights become operationalized through user organizations. Ensuring that women, smallholders, livestock keepers, or other poor and marginalized water users are represented in these organizations is an important step to strengthening their water rights, but is often difficult, because of overt resistance from those who do not want to share water rights and decision-making, or because of social challenges of including marginalized groups in local organizations.

Officially-recognized rights also help ensure that their holders have a "seat at the table" in discussions about further water development or land use changes that may impinge on their rights. Joint planning and modeling of water resource development with government agencies and different user groups helps to put this into practice, but it may require strengthening the capacity of both the agencies and the users. There is no single, optimal property right system for irrigations systems—in developing countries or elsewhere. Rather, we need a range of options and the understanding necessary to be able to tailor them to their (ever-changing) physical and institutional context.

ACKNOWLEDGMENTS

This paper draws on the framework of the CGIAR Program on Collective Action and Property Rights (CAPRi), which is part of the CGIAR Research Program on Policies, Institutions and Markets. Special thanks are due to Henning Bjornlund Sarah Wheeler for organizing the conference that led to this special issue, for the invitation to be part of it, and for their insightful comments, encouragement and patience.

REFERENCES

1. Alchian and Demsetz, 1972 A. Alchian, H. Demsetz Production, information costs, and economic organization Am. Econ. Rev., 62 (5) (1972), pp. 777–795.
2. Arnold, 2002 C.A. Arnold The reconstitution of property: property as a web of interests Harvard Environ. Law Rev., 26 (2) (2002), p. 281.
3. Asquith, 2006 N. Asquith Bees and barbed wire on the Bolivian frontier PERC Rep., 24 (4) (2006), pp. 3–6.
4. Bauer, 2004 C.J. Bauer Siren Song: Chilean Water Law as a Model for International Reform Resources for the Future, Washington, DC (2004).

5. Bjornlund and McKay, 2002 H. Bjornlund, J. McKay Aspects of water markets for developing countries: experiences from Australia, Chile and the US Environ. Dev. Econ., 7 (4) (2002), pp. 767–793.

6. Blackstone, 1766 Blackstone Commentaries on the Laws of England (1766). Book II (1766) <http://avalon.law.yale.edu/18th_century/blackstone_bk2ch2.asp> (accessed 11 Aug 2013).

7. Boelens and Hoogendam, 2002 R. Boelens, P. Hoogendam (Eds.), Water Rights and Empowerment, Van Gorcum, Assen, The Netherlands (2002).

8. Bromley, 1991 D. Bromley Environment and Economy: Property and Policy Basil Blackwell, Cambridge, MA (1991).

9. Bruns et al., 2005 B.R. Bruns, C. Ringler, R.S. Meinzen-Dick (Eds.), Water Rights Reform: Lessons for Institutional Design, International Food Policy Research Institute, Washington, DC (2005).

10. Bues and Theesfeld, 2012 A. Bues, I. Theesfeld Water grabbing and the role of power: shifting water governance in the light of agricultural foreign direct investment Water Altern., 5 (2) (2012), pp. 266–283.

11. Chong and Sunding, 2006 H. Chong, D. Sunding Water markets and trading Annu. Rev. Environ. Resour., 31 (2006), pp. 239–264.

12. Cleaver, 2001 F. Cleaver Institutional bricolage, conflict and cooperation in Usangu IDS Bull., 32 (4) (2001), pp. 26–35.

13. Commons, 1968 J.R. Commons Legal Foundations of Capitalism University of Wisconsin Press, Madison, WI (1968).

14. Coward, 1986 E.W. Coward Jr. Direct or indirect alternatives for irrigation investment and the creation of property K.W. Easter (Ed.), Irrigation Investment, Technology and Management Strategies for Development, Westview Press, Boulder, CO (1986).

15. Cox et al., 2010 M. Cox, G. Arnold, S. Villamayor Tomás A review of design principles for community-based natural resource management Ecol. Soc., 15 (4) (2010), p. 38.

16. Deininger, 2003 K. Deininger Land Policies for Growth and Poverty Reduction: A World Bank Policy Research Report Oxford University Press, New York, NY (2003).

17. Easter and Huang, in press Easter K.W., Huang, Q. (Eds.), in press. Water Markets for the 21st Century: What We Have Learned? Springer; Dordrecht, the Netherlands.

18. Easter et al., 1998 K.W. Easter, M.W. Rosegrant, A. Dinar (Eds.), Markets for Water: Potential and Performance, Kluwer, Boston, MA (1998).

19. Eggertsson, 1990T. Eggertsson Economic Behavior and Institutions Cambridge University Press, Cambridge, U.K (1990).

20. Ehrlich, 1936 E. Ehrlich Fundamental Principles of the Sociology of Law Harvard University Press, Cambridge, MA (1936).

21. FAO, 2012 FAO (Food and Agriculture Organization of the United Nations) Voluntary Guidelines on the Responsible Governance of Tenure of Land, Fisheries and Forests in the Context of National Food Security FAO, Rome (2012).

22. Faruqui et al., 2001 N.I. Faruqui, A.K. Biswas, M.J. Bino (Eds.), Water Management in Islam, United Nations University Press and International Development Research Centre, Tokyo (2001).

23. Furubotn and Pejovich, 1972 E.G. Furubotn, S. Pejovich Property rights and economic theory: a survey of recent literature J. Econ. Lit., 10 (4) (1972), pp. 1137–1162.

24. Garces-Restrepo et al., 2007 C. Garces-Restrepo, D. Vermillion, G. Muñoz Irrigation Management Transfer International Water Management Institute and FAO, Rome (2007).

25. Hodgson, 2004 S. Hodgson Land and Water—The Rights Interface FAO Legislative Study 84FAO, Rome (2004).

26. Hodgson, 2013 S. Hodgson Water rights and water tenure Background Paper 3 for FAO Expert Consultation on Water Governance and the Role of Tenure and Rights in Coping with Agricultural Water ScarcityFAO, Rome (2013), pp. 21–23.

27. Ingram and Brown, 1998 H. Ingram, F.L. Brown Commodity and community water values. Experiences from the U.S. Southwest R. Boelens, G. Davila (Eds.), Searching for Equity: Conceptions of Justice and Equity in Peasant Irrigation, Van Gorcum, Assen, The Netherlands (1998), pp. 114–120.

28. Kemerink et al., 2011 J.S. Kemerink, R. Ahlers, P. van der Zaag Contested water rights in post-apartheid South Africa: the struggle

for water at catchment level Water SA, 37 (4) (2011), pp. 585–594.

29. Komakech et al., 2012 H.C. Komakech, M. Condon, P. van der Zaag The role of statutory and local rules in allocating water between large- and small-scale irrigators in an African river catchment Water SA, 38 (1) (2012), pp. 115–126.
30. Komakech and van der Zaag, 2013 H.C. Komakech, P. van der Zaag Polycentrism and pitfalls: the formation of water users forums in the Kikuletwa catchment, Tanzania Water Int., 38 (3) (2013), pp. 231–249.
31. Knox et al., 2002 A. Knox, R. Meinzen-Dick, P. Hazell Property rights, collective action, and technologies for natural resource management: a conceptual framework R.S. Meinzen-Dick, A. Knox, F. Place, B.M. Swallow (Eds.), Innovation in Natural Resource Management: The Role of Property Rights and Collective Action in Developing Countries, Johns Hopkins University Press, Baltimore, MD (2002), pp. 12–44.
32. Lam, 1998 W.F. Lam Governing Irrigation Systems in Nepal: Institutions, Infrastructure, and Collective Action ICS Press, San Francisco, CA (1998).
33. Lansing, 1991J.S. Lansing Priests and Programmers: Technologies of Power in the Engineered Landscape of Bali Princeton University Press, Princeton, NJ (1991).
34. Libecap, 1998 G. Libecap Distributional and political issues in modifying traditional common-property institutions E. Berge, N.C. Stenseth (Eds.), Law and the Governance of Renewable Resources: Studies from Northern Europe and Africa, ICS Press, Oakland CA (1998), pp. 75–92.
35. Meinzen-Dick, 2002 R.S. Meinzen-Dick Property rights and maintenance of irrigation systems Z. Bewässserungswirtschaft (J. Appl. Irrig. Sci.), 37 (2) (2002), pp. 219–236.
36. Meinzen-Dick, 1996 R. Meinzen-Dick Groundwater Markets in Pakistan: Participation and Productivity International Food Policy Research Institute, Washington, DC (1996), p. 105 Research Report.
37. Meinzen-Dick and Nkonya, 2007 R. Meinzen-Dick, L. Nkonya Understanding legal pluralism in water and land rights: lessons

from Africa and Asia B. van Koppen, M. Giordano, J. Butterworth (Eds.), Community-based Water Law and Water Resources Management Reform in Developing Countries, Comprehensive Assessment of Water Management in Agriculture Series 5, CAB International, Wallingford, UK (2007), pp. 12–27.

38. Meinzen-Dick and Pradhan, 2002 R.S. Meinzen-Dick, R. Pradhan Legal pluralism and dynamic property rights CAPRi Working Paper No. 22International Food Policy Research Institute, Washington, DC (2002).

39. Merrey et al., 2007 D.J. Merrey, R.S. Meinzen-Dick, P.P. Mollinga, E. Karar Policy and institutional reform processes for sustainable agricultural water management: the art of the possible D. Molden (Ed.), Water for Food, Water for Life, Earthscan, London (2007), pp. 193–232.

40. North, 1990 D.C. North Institutions, Institutional Change and Economic Performance Cambridge University Press, Cambridge, UK (1990).

41. Onyango et al., 2007 L. Onyango, B. Swallow, J.L. Roy, R. Meinzen-Dick Coping with history and hydrology: how Kenya's settlement and land tenure patterns shape contemporary water rights and gender relations in water B. van Koppen, M. Giordano, J. Butterworth (Eds.), Community-Based Water Law and Water Resources Management Reform in Developing Countries, CABI, London (2007), pp. 173–195.

42. Ostrom, 1990 E. Ostrom Governing the Commons: The Evolution of Institutions for Collective Action Cambridge University Press, New York, NY (1990).

43. Ostrom, 1992 E. Ostrom Crafting Institutions for Self-Governing Irrigation Systems Institute for Contemporary Studies, San Francisco, CA (1992).

44. Otsuka and Place, 2001 K. Otsuka, F. Place Issues and theoretical framework K. Otsuka, F. Place (Eds.), Land Tenure and Natural Resource Management: A Comparative Study of Agrarian Communities in Asia and Africa, Johns Hopkins University Press, Baltimore, MD (2001), pp. 3–33.

45. Pandolfelli et al., 2008 L. Pandolfelli, R. Meinzen-Dick, S. Dohrn Introduction: gender and collective action: motivations, effectiveness and impact J. Int. Dev., 20 (1) (2008), pp. 1–11.

46. Pradhan and Pradhan, 2000 R. Pradhan, U. Pradhan Negotiating access and rights: disputes over rights to an irrigation water source in Nepal B.R. Bruns, R.S. Meinzen-Dick (Eds.), Negotiating Water Rights, Intermediate Technology Press, London (2000), pp. 200–221.
47. Ribot and Peluso, 2003 J. Ribot, N.L. Peluso A theory of access Rural Sociol., 68 (2) (2003), pp. 153–181.
48. Rogers et al., 1995 P. Rogers, R. Bhatia, A. Huber-Lee Water as a social and economic good TAC Background Papers No. 2Global Water Partnership, Stockholm (1995).
49. Rosegrant and Binswanger, 1994 M.W. Rosegrant, H.P. Binswanger Markets in tradable water rights: potential for efficiency gains in developing country water resource allocation World Dev., 22 (11) (1994), pp. 1613–1625.
50. RSA, 1998 RSA (Republic of South Africa), 1998. *National Water Act.* Gov. Gaz. 398, No. 19182, Cape Town, South Africa.
51. Schlager and Ostrom, 1992 E. Schlager, E. Ostrom Property rights regimes and natural resources: a conceptual analysis Land Econ., 68 (3) (1992), pp. 249–262.
52. Schreiner, 2013 B. Schreiner Viewpoint—why has the South African National Water Act been so difficult to implement? Water Altern., 6 (2) (2013), pp. 239–245.
53. Scott, 2008 A. Scott The Evolution of Resource Property Rights Oxford University Press, Oxford, UK (2008).
54. Shah and Raju, 1988 T. Shah, K.V. Raju Ground water markets and small farmer development Econ. Political Weekly, 23 (13) (1988), pp. A23–A28.
55. Solanes, 2013 M. Solanes Viewpoint—The Washington Consensus, Chilean water monopolization and the Peruvian draft water law of the 1990 Water Altern., 6 (2) (2013), pp. 207–217.
56. Swallow et al., 2002 B. Swallow, D.P. Garrity, M. Van Noordwijk The effects of scales, flows and filters on property rights and collective action in watershed management Water Policy, 3 (6) (2002), pp. 457–474.
57. Swallow et al., 2006 B. Swallow, N. Johnson, R. Meinzen-Dick, A. Knox The challenges of inclusive cross-scale collective action in watersheds Water Int., 31 (3) (2006), pp. 361–375.

58. Tang, 1992S. Y. Tang Institutions and Collective Action: Self-governance in Irrigation ICS Press, San Francisco, CA (1992).
59. Uphoff, 1986 N. Uphoff Improving International Irrigation Management with Farmer Participation: Getting the Process Right Westview Press, Boulder CO (1986).
60. van Koppen et al., 2007 B. van Koppen, M. Giordano, J. Butterworth (Eds.), Community-Based Water Law and Water Resources Management Reform in Developing Countries, CABI, London (2007).
61. Vermillion, 1987 D.L. Vermillion Turning over irrigation systems from the government of Indonesia to farmers IIMI Working Paper 2International Irrigation Management Institute, Digana Village, Sri Lanka (1987).
62. Vermillion, 1997 D.L. Vermillion Impacts of irrigation management transfer: a review of the evidence Research Report 11International Irrigation Management Institute, Colombo, Sri Lanka (1997).
63. von Benda-Beckmann et al., 1998 F. von Benda-Beckmann, K. von Benda-Beckmann, J. Spiertz Equity and legal pluralism: taking customary law into account in natural resource policies R. Boelens, G. Davila (Eds.), Searching for Equity: Conceptions of Justice and Equity in Peasant Irrigation, Van Gorcum, Assen, The Netherlands (1998), pp. 57–69.
64. Young, in pressm Young, M., in press. Designing water entitlement regimes for an ever-changing and ever-varying future. Agric. Water Manage, http://dx.doi.org/10.1016/j.agwat.2013.12.002.
65. Zekri and Al-Marshudi, 2008 S. Zekri, A.S. Al-Marshudi A millenarian water rights system and water markets in Oman Water Int., 33 (3) (2008), pp. 350–360.

Chapter 3

Irrigation Management in Coastal Zones to Prevent Soil and Groundwater Salinization

Nicolas Greggio, Pauline Mollema, Marco Antonellini and Giovanni Gabbianelli

[1]Interdepartmental Centre for Environmental Science Research (C.I.R.S.A.), University of Bologna, Ravenna,, Italy

INTRODUCTION

Soil salinization is one of the most widespread soil degradation processes on earth and, worldwide, one billion hectares are affected, mainly in the arid–semiarid regions of Asia, Australia and South America [1]. In Europe, soil salinity has effects on one million hectares mainly in the Mediterranean countries [1]. There are two types of salinization: primary salinization caused by natural events such as sea spray or rock weathering or seepage [2] and secondary salinization that is caused by human activities such as irrigation with salty water, groundwater overexploitation and excessive drainage [1].

Along the Adriatic coast of the Po Plain, freshwater resources are becoming increasingly scarce, because of irrigation and other intense water use, salinization and long periods of drought [3]. Custodio [4] underlines that, especially in southern Europe, the irrigation practices and the water requirements to sustain the coastal tourism industry exhort a strong pressure on water resources.

The impact of groundwater salinization in coastal areas affects both natural vegetation biodiversity and agricultural production, through soil salinization and reduction of freshwater availability for irrigation. Salinization is closely associated with the process of desertification, because salinity may have direct negative effects on crop yields by reducing the ability of plant roots to take up water [5]. The most common salinity effect is a general stunting of plant growth, but not all plants respond in the same way. Grain and corn may reduce their seed production without appreciably plant dimensions reduction. Wheat, barley, cottons seed productions, on the other hand, are decreased less than their vegetative growth [6]. The only agronomical significant criterion for evaluating soil salinization effects is the commercial crop yield decrease [6]. This encourages farmers in certain areas to cultivate salt resistant crops to adapt to high salinity of the soil and groundwater [5]. In our area fresh irrigation water until now was readily available and no salt resistant crops are grown, or other adaptation measures were taken against soil and groundwater salinization of the farmland.

An increase in water and soil salinity within a coastal habitat has direct effects on the natural vegetation and agriculture [5],[7], including decrease in plant species richness, decrease in wetland dry biomass, changes in plant communities, decrease in seedling germination, decrease in surface area of the leaves, decrease in stem length, N2-fixation inhibition, increasing mortality, and indirect effects such as habitat loss for some animal species [8],[9],[10].

In the Ravenna province, more than 68 square kilometers of farmland are at a risk for soil salinization (Figure 3). These are primarily the areas near the Pialasse lagoons and near the rivers open to sea (Figure 1). This agricultural land is at sea level and drained by pumping machines. Many of the pumping machines are located 5 km far from the shoreline and they maintain the water table at 2 meters below sea level during the year, creating hydraulic gradients land inwards and promoting salt-qwater intrusion from the Adriatic Sea [3].

High rates of anthropogenic and natural subsidence and artificial drainage, among others causes, have caused groundwater salinization in the coastal unconfined and semi-confined aquifer near Ravenna, [3],[11] and a subsequent loss of fertile soils and vegetation species richness in the natural areas [7].

In the Ravenna Municipality just about 35000 hectares are agricultural land, of these 30000 are arable land and 5000 are orchards. Irrigation is widely diffuse and more than 50% of farms are equipped for irrigation. Until two decades ago the main source of irrigation water was groundwater from the phreatic and confined aquifers, but the increase of subsidence rates, forced to change the water supply from groundwater to surface water [12],[13], mostly from the Po River and some from the minor rivers flowing from the Apennines.

The phreatic aquifer consists of sandy beach and dune deposits and is unconfined close to the sea and confined by a clay layer further inland (Figures 11and 12). Two dune belts parallel to the coast covered by pine forests form the only topography above sea level. The area in between the dunes belts is used for irrigated and rain-fed farmland, as well as the land behind the older dune belt, 5 km from the coast (Figure 1).

Most of the coastal aquifer near Ravenna is very salty (above 10 g/l) [3][11][14] but, besides some small shallow freshwater lenses at the top of the aquifer, we found freshwater below an irrigated field. In some ways this result is worrying, because irrigation water should not reach the groundwater, so that leaching of nitrates is avoided and waste of resources prevented. Mollema et al. ([15]) concluded from a water budget analysis over the *Quinto* basin watershed (south of the Ravenna urban area) that the amount of water used for irrigation was abundant but only looked at total volumes for the whole basin and did not quantify the contribution of irrigation to groundwater.

The objective of this study is to quantify in detail the effect of irrigation water on the groundwater hydrology with the help of geochemical analysis, geophysical profiling and infiltration measurements. We address, in particular, the mechanism of how excess irrigation water ends up in the aquifer and how that affects the salinity of the groundwater. By doing all this, we assess whether irrigation practices can help to counteract further groundwater and soil salinization.

STUDY AREA

The study area (red rectangle in Figure 1) is part of the *Quinto* basin, the watershed that is confined between two rivers flowing from the Northern Apennines to the Adriatic Sea: the *Uniti* River in the north and the *Bevano* River in the south. The eastern border of the basin is the Adriatic Sea and the western border follows, in part, the *Ronco* River, a tributary of the *Uniti* River.

An east-west profile through the area starting at the shoreline sea includes a beach near the town of*Lido di Dante*, a recent dune belt covered by a coastal pine forest called *Ramazzotti*, planted at the beginning of the 1900's, agricultural land, an older dune belt (paleodunes) covered with a pine forest (*Classe* pine forest), a zone with many active and abandoned gravel quarries that form lakes and more agricultural land (Figures 1 and 5).

Our focus is on a plot of agricultural area (500 by 500 m) in the eastern part of the profile (Figure 2). The area is around a ditch (shown in red). The green dots are wells used to monitor the freshwater lenses, in particular P1S, P2S and P3S are fully penetrating the aquifer (-27m), the others are shallow piezometers (-6m).

The ditch was dug in 1981 as a reservoir for irrigation water that was withdrawn from the *Fiumi Uniti*River. In 2008, the supply for irrigation changed and now it comes through an underground pipe from the *Po* River. The ditch is 500 m long, 5 m wide and 2,5m deep, resulting in a total volume of 6250 m^3; it is completely disconnected from the drainage channels network.

In 2011, the planting schedule was as follows: in the east parcel tomato plants were grown from April to August; Alfa Alfa was contemporaneously grown in the west parcel. Because of the particularly dry 2011 weather conditions, tomato plants needed irrigation water from the early stage of growth onward.

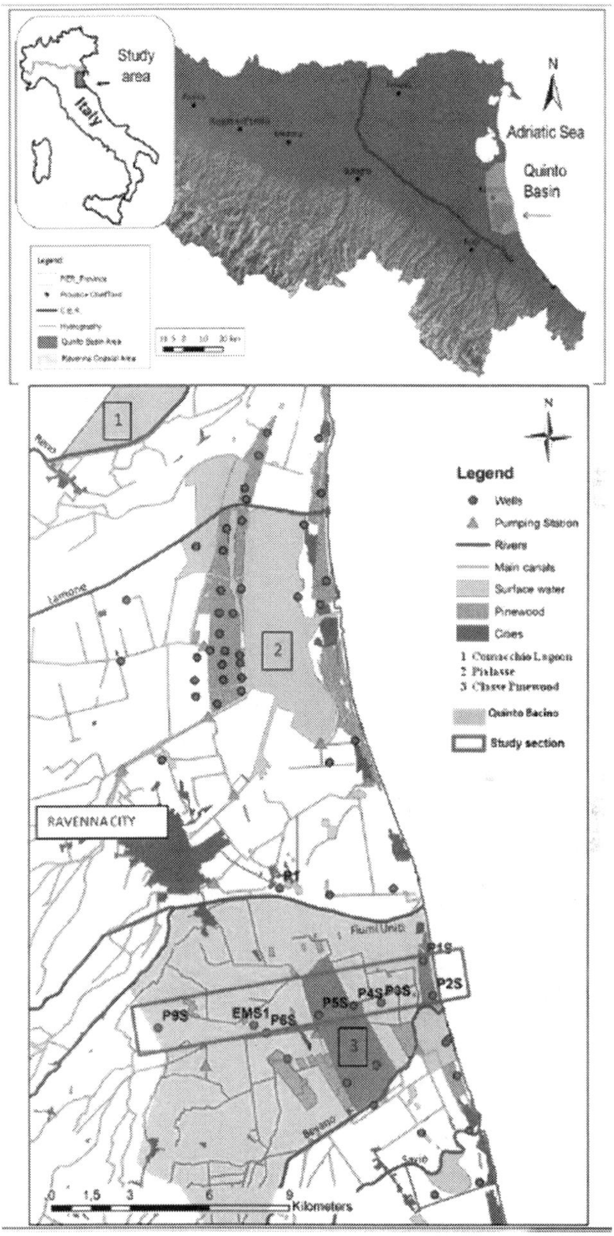

Figure 1. Index map and detail of the Ravenna coastal zone and location of study area (*Quinto* basin); the map reports also monitoring wells, rivers, channels, pumping stations and main environmental features.

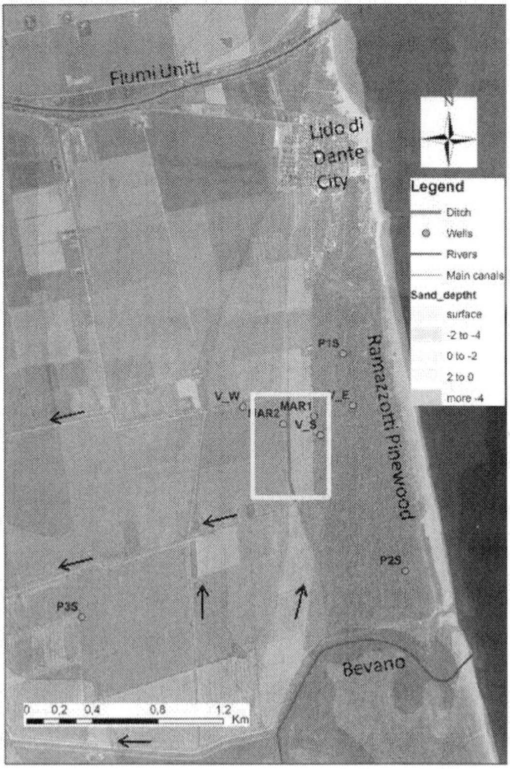

Figure 2. Detail of the study area with thickness of sandy top layer, main drainage channels, flux direction and wells used for water table monitoring.

SOIL TYPE AND LAND USE

In the coastal area, sediments at the surface are mainly sandy (Figure 3). The sand layers are parallel to the coastline and form paleo and recent dunes.

The sandy soils are called *Cerba* soils, and according to the FAO classification [16] for Soil Taxonomy, they are Calcaric Arenosols and Mesic Aquic Ustipsamments [17].

These soil types are mainly used as arable land or horticulture. Due to their high hydraulic conductivity, these sandy soils are most sensitive for salinization [18],[19]. The *Sant'Omobono* soils (Brown in Figure

3) represents river sand deposits; these soils were deposited along recent or paleo-rivers during flooding events. According to the FAO classification [16] for the Soil Taxonomy they are Calcaric Cambisols, Fine Sandy, Mixed, mesic Udifluventic Ustochrepts, [17]. These kinds of soils are cultivated with vineyards and orchards. Because of their sandy texture, also these soils are at risk of salinization. The *Medicina* (*Cataldi*) sediments (Green colors) are soils with fine texture, mainly created when rivers flowed in presently reclaimed areas. FAO classifies them as Eutric Vertisols [16] and for Soil Taxonomy they are Fine, Mixed, Mesic Ustic Endoaquerts. [17]. They are grown with arable crops only and they are not affected by salinization [18],[19],[20].

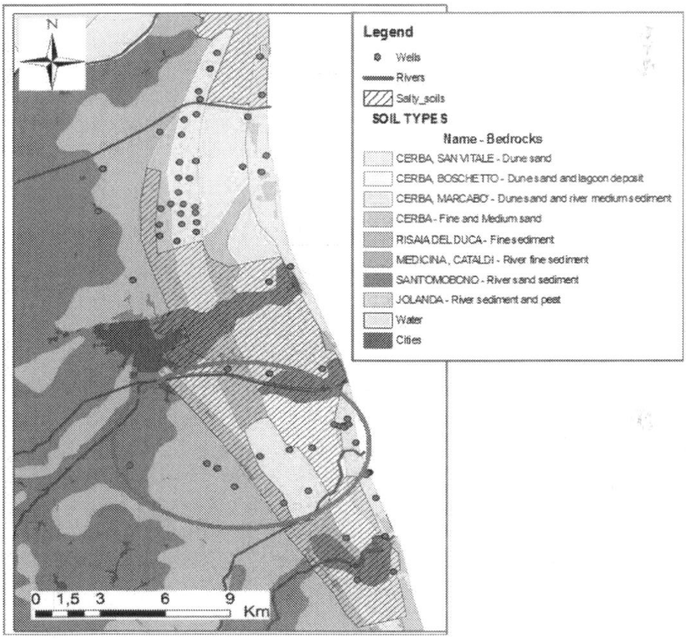

Figure 3. Soil types in the Ravenna area and soils at risk of salinization.

The area studied in detail has an extension of 12 ha consisting of two parcels of land of 6 ha on each side of the ditch. The western parcel lies on a 1-m-thick silty sand top layer (*Cerba Boschetto* soil) that confines the sandy aquifer below, whereas the eastern parcel lies on the unconfined part of the aquifer that consists of beach sand (*Cerba San Vitale* soil; Figures 2 and 4).

The differences in soil and lithology, make the land suitable for different kinds of crops: in the last thirty years, on the sandy soil (eastern parcel), the crops consisted of horticulture (strawberries, tomatoes, potatoes, nursery sugarbeet); on the silty soil instead (western parcel), typical extensive arable crops were grown (barley, wheat, maize, alfa-alfa). At our Mediterranean latitude, horticulture crops need irrigation during all the growth phases [13]; arable crops, on the other hand, do not need irrigation during growth. To provide irrigation water for horticulture, the ditch is kept full of freshwater in the summer period. As a consequence, the water level in the ditch is higher than the water table level in the adjacent aquifer. During the winter, the water level in the irrigation ditch matches the adjacent water table level.

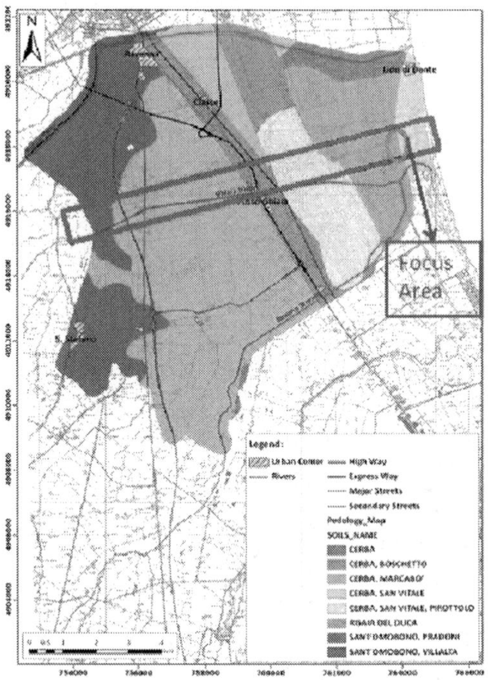

Figure 4. *Quinto* Basin soil types, with study section and focus area.

Most of the land in the coastal zone near Ravenna is used for agriculture but urban and natural areas (pinewood and water surface body) are also widespread (Figure 1). Most of the natural environments are part of the Po Delta Regional Park. Irrigable crops are the most occurring culture in the province of Ravenna. Vineyards and orchards occupy a small percentage of the territory mostly further inland [12],[13].

The *Quinto* Basin covers a total of 10,355 hectares of which the largest part is used for agriculture (66%), 10% is covered by pine forests, 9% is urban area, and 9% is open surface water or wetlands (Figure 5) [15],[21].

The agricultural land is primarily devoted to arable crops as wheat, barley, alfa alfa, sorghum and corn and only 15% is grown with horticultural and orchards. Horticulture represents less than 5% of the*Quinto* basin and it is practiced on sandy soils (*Cerba* soil) (Figure 4). Orchards are grown in the western part, on *Sant'omobono* soils only that, being the deeper soils in the *Quinto* basin, are the most adapt for fruits and grapes, which are sensitive to water logging (Figures 4 and 5). The most common horticultural products cultivated in the *Quinto* basin are tomatoes, potatoes and nurseries of strawberries and sugar beets. The dominant fruit trees are peaches, apples and vine. For this basin, the data from the last general agricultural census shows that the entire surface cultivated with fruits and horticultural crops are equipped with irrigation systems [12].

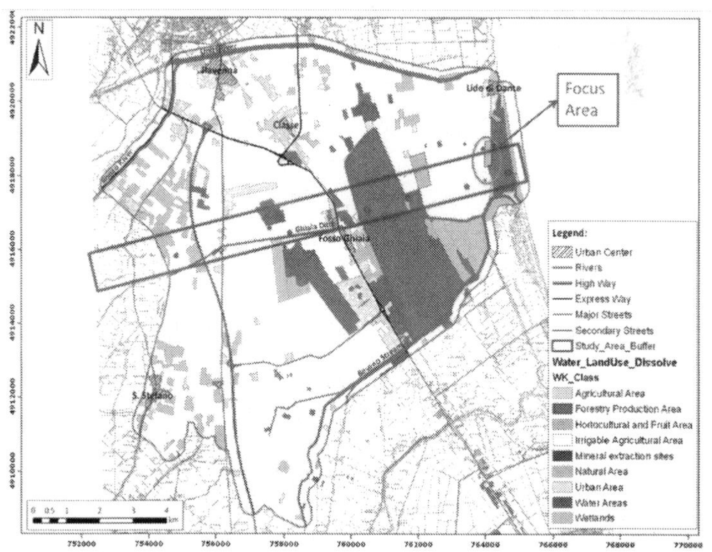

Figure 5. *Quinto* Basin land use map.

TOPOGRAPHY, DRAINAGE AND IRRIGATION

Most of the land near Ravenna is below or just a few centimeters above sea level. Only the recent dune belt and the paleo-dunes are up to 5 m above sea level (Figure 6). These dunes are covered by pine forests (Figure 1). The maximum elevation in the *Quinto* Basin is 5 m and it is measured in the most recent dune belts, whereas the lowest areas are near the gravel quarries 7 km from the coast at 2m below sea level (Figure 6). The farmland around the ditch of our study area is completely flat and lies at a few centimeters above sea level.

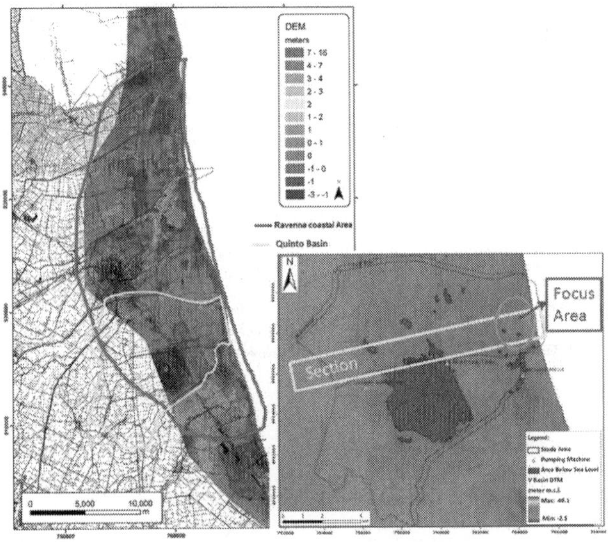

Figure 6. Topography of the coastal zone near Ravenna and area below sea level in the *Quinto* basin.

The whole coastal zone of Ravenna is drained mechanically to make agriculture possible despite the low topography. As a consequence, the areas of natural water infiltration are very small and localized along the dune belts. The entire district is divided into small watersheds defined by small topographic features such as rows of dunes, dikes, and levees [22]. Figure 7 shows the drainage basins in the Ravenna Municipality. Only one basin has a natural hydraulic gradient towards the sea (Figure 7).

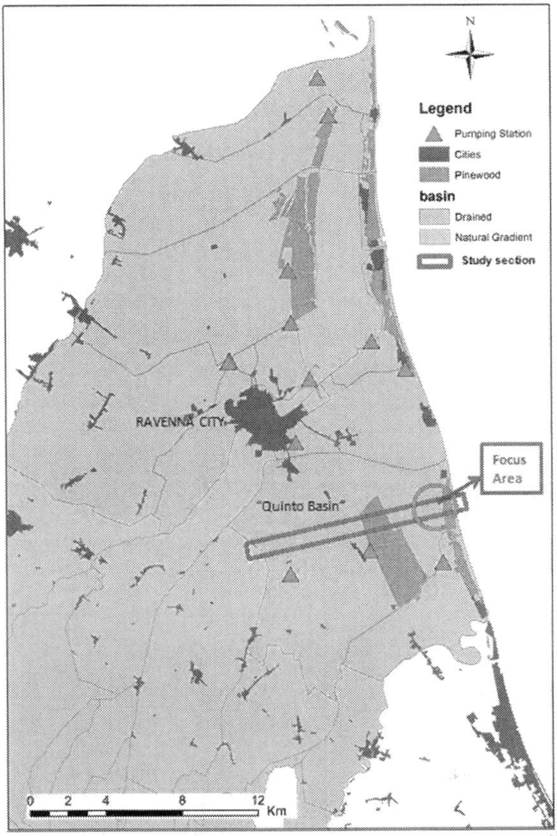

Figure 7. Drainage basins in the Ravenna Municipality. The light green color represents natural drainage basins; the pink color shows the mechanically drained basins.

A dense network of pumping stations and channels was built in the last century with the aim to keep the land from flooding and allow farming [23]. In the *Quinto* basin alone, more than 50 km of drainage channels reroute the water to the pumping stations [24].

The pumping stations are typically located inland, several kilometers from the shoreline (Figure 7). The drainage, therefore, creates a hydraulic gradient directed inland [3],[23]. Our study area is drained by two east-west oriented channels that collect the water between the two dune belts and convey it towards the *Fosso Ghiaia* pumping station (Figure 2).

The pumping stations are activated automatically when the water level in the drainage channels reaches a threshold, regardless of the soil water availability or the weather forecasts. As a consequence, the water table is everywhere below sea level with the exception of the dune belts where the water table may be above sea level in the wet periods of the year. The pumping station in the *Quinto* basin, (*Fosso Ghiaia*) causes a deep withdrawal cone down to -2.5 masl in the western part of the *Classe* pinewood (Figure 8). The flux vectors that were calculated based on water table measurements in the wells ofFigure 1 are all directed inland showing that sea water flows into the aquifer.

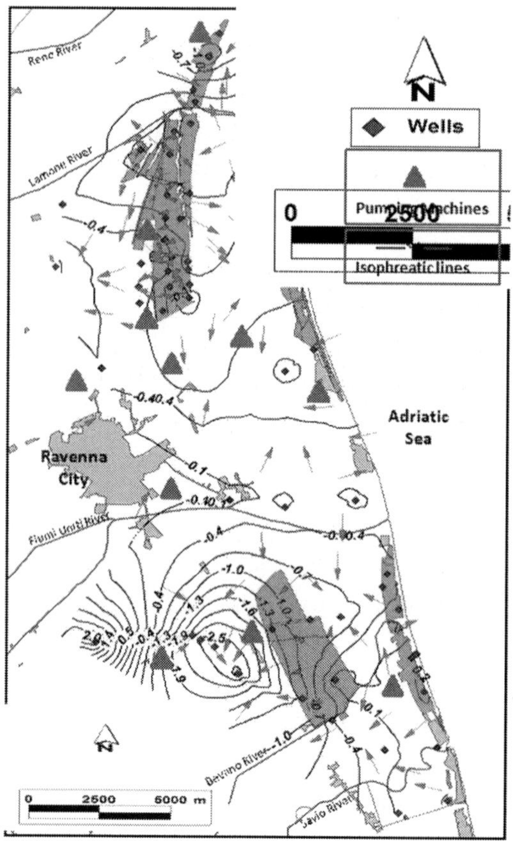

Figure 8. Water table contour map and flux vectors indicating the direction of groundwater flow.

The main source of irrigation water is from river water; five rivers are tapped for irrigation, from north to south they are the *Reno, Lamone, Fiumi Uniti, Bevano* and *Savio* rivers. In the summer period, however, there is not enough water in these rivers and additional irrigation water is taken from the *Po* River by means of the *Canale Emiliano Romagnolo* channel (CER). The water from the CER is brought into the irrigation systems via the same channels that are used for drainage in the winter.

The most common irrigation systems are the sprinkler and the drip irrigation system. The first one is used for horticultural crops with a large volume of water distributed twice a week, whereas the second one is typical for orchards and it uses small volumes of water but with a daily treatment. The sprinkler method, therefore, needs a storage basin for the water (ditch, lake or basin). The drip system instead doesn't need a storage basin [25].

In our focus area, water for irrigation was taken for many years from the *Fiumi Uniti* River. Seawater, however, has encroached along the river and, therefore, it was necessary to change the water supply [3]. Recently, irrigation water comes through a pressurized pipeline directly from the CER (Po river source). During the monitoring time, irrigation on tomato parcel started at an early stage of growth in April 2011 and ended at the harvest time in August 2011. The scheduling of irrigation has been rather regular: approximately twice a week. For each irrigation operation, 20 mm/m^2 were distributed on the tomato plants (Table 1). No irrigation was given to the alfa alfa in the parcel to the west of the ditch.

In the Ravenna Municipality, irrigation and drainage practices are managed by the same authority: the Land Reclamation Consortium for the *Romagna* Region and for both services the end users costs have recently increased [24],[25].

CLIMATE

The climate of Ravenna is Mediterranean with generally warm summers and mild winters [26]. The average temperature is about 14°C varying from -5° to 10°C in the winter period and from 20° to 35° C in the summer months [27]. The average annual rainfall calculated on a long term database (1960-2011) is more than 650 mm and it is usually concentrated in the spring and autumn months.

In five of the last six years, total precipitation has been under the average; in particular 2011 was the driest of the last 23 years, with only 385 mm (Figure 9). Compared with the average monthly rainfall of 1960-2011, in fact, most of the autumn rainfall is absent (Figure 9). The same comparison for temperature shows that in 2011 the temperature was 3-4°C higher than the average of the last 51 years, especially in autumn and winter (Figure 10).

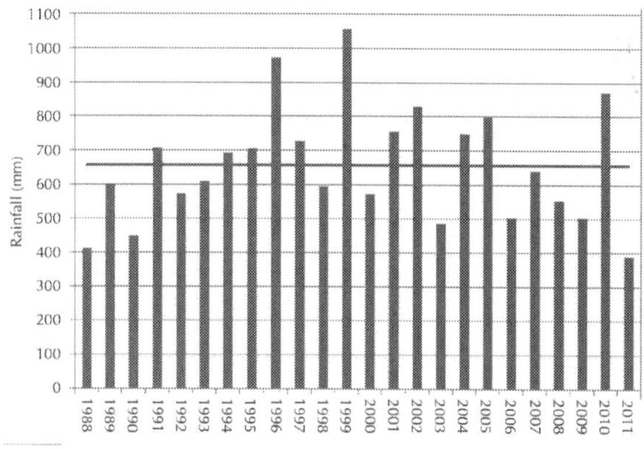

Figure 9. Annual precipitation in the city of Ravenna. Red line represents the long term mean value.

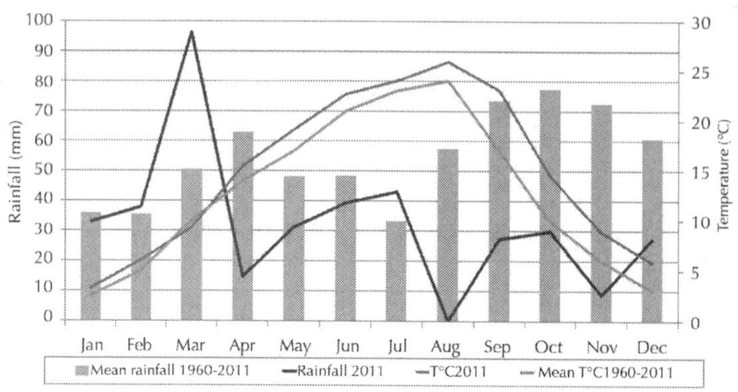

Figure 10. Monthly rainfall and temperature for the Ravenna Area.

One of the climate change scenarios proposed for the Mediterranean area [28] indicates a substantial change in the hydrologic budget by the end of this century. According to Antonellini and Mollema 2010 ([7]), evapotranspiration will increase as a result of an increase in temperature; rainfall will be concentrated in short periods, the winter will become more humid and the summer more dry compared to the present, and there will be an increase in extreme events (floods or droughts). Such a forecast for the coastal zone has a particular significance, because a shortage of fresh water in the area next to the sea will promote saltwater intrusion in the ground water and soil salinization. According to Magnan et al. 2009 ([28]), it is possible to mitigate some of these effects, but it is not possible to avoid them completely; it is, therefore, necessary to start the development of mitigation and adaptation strategies now in order to face these issues.

GEOLOGY AND GEOMORPHOLOGY

The study area is a small portion of the southern part of the wider Po delta plain. The surface aquifer sediments were deposited during the Late Quaternary (Holocene) sea-level fluctuations [29], [30]. During the last glacial maximum (19 kyr ago) the sea-level dropped about 100-120 m in the Adriatic basin and the North Adriatic Sea was became a deep alluvial plain [31] [32]. During the optimum climatic period, about 5500 years ago, the paleo-shoreline was located 25-30 km west of its actual position. From the most landward position of the Holocene coast, a rapid migration of a barrier-lagoon-estuarine system was developed to create the Ravenna alluvial plain, especially from Roman age (IV cent. B.C.) to recent times.

The sedimentation in the study area turned from marine to continental only recently (XIX century). Most of the study area emerged during the last shoreline progradational phase [33], which took place from 1851 to 1894 A.D at an average rate of 20 m/y and in the first half of the past century (4-12 m/y). Nowadays, although the Holocene glacial period with related sea level changes is finished [34], the movements of the coastline still occur: a large part of the coast is under active erosion and only a small part is prograding to sea. This is mostly due to human intervention: the subsidence due to water and gas exploration in combination with a lack of sediments transport by the rivers and the construction of various piers perpendicular to the coast near the harbors and breakwaters parallel to the coast that trap the sediments. These interventions have interrupted the natural process of long shore current and sediment transport [34]. Now beach and dune erosion and shoreline retreat are a major problem along the Adriatic coast [35].

The sandy aquifer consists of two main units: a relatively thick medium-grained sand unit (from 0 to -10 masl) and a lower fine-grained sand unit of small thickness (from -21 to -26 masl). These two sand bodies are separated by a clay-silt and sand-silt continental unit (from -10 m to -21 masl) called*Prodelta* (Figure 11). Lastly, the Flandrian continental silty-clay basement is at a depth varying from -20 m.a.s.l. in the west to -30 m.a.s.l. at the present shoreline [29], [36],[37].

The resulting picture of the aquifer is a wedge shaped dune and beach sand body pinching out in a westerly direction. It is enclosed at the bottom and top by clays and peat deposited in lagoons, marshes and alluvial plains. The only portion where the aquifer is phreatic are the paleodunes and the actual dune belt both covered by pine forests. At the tip of this wedge there are gravel deposits (Figure 11) [29] [38].

Figure 11. Lithological cross section of the Ravenna aquifer in the study area (modified from [29] [39]).

MONITORING

A groundwater monitoring campaign has been carried out in 8 wells along the section of Figure 1, in December 2010 and it was repeated in

April 2011, with the aim to compare salinity and chemical data in two different seasons. A monthly water table monitoring was performed for the whole transect with increased data density around the ditch, where we installed two shallow wells in February 2011. Overall, 8 wells near the ditch were monitored for water table analysis (Figure 15).

Rainfall was measured with 5 pluviometers located along the section. In the eastern parcel, within the tomato field, we were able to measure both rainfall and irrigation. To assess how much rainfall and irrigation from the surface reaches the groundwater, we buried in the same location a drain-gauge. Geo-electric vertical resistivity surveys (VES) in cross sections normal to the ditch (Figure 12) were carried out in order to determine the depth, shape, and evolution of the fresh water lens [40],[41],[42]. VES surveys were repeated twice: before the tomatoes were planted in February 2011 and during the maximum irrigation period in July 2011. The VES calibration was made using well salinity data, and apparent resistivity data obtained for the area by the process of inversion [43] [44].

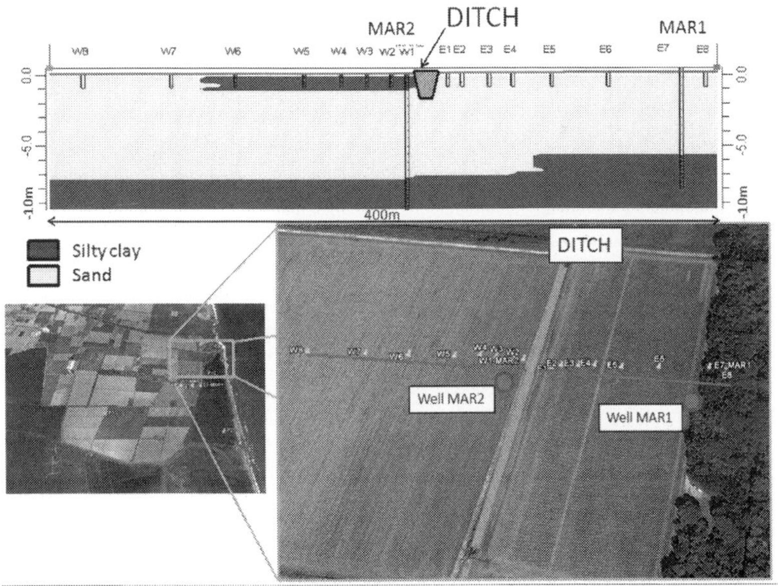

Figure 12. Small parcel, studied in detail, with simplified lithology, irrigation ditch (light blue), wells MAR1 and MAR2 (red circles), and 16 V.E.S. locations.

Groundwater with a salinity ranging from 0 to 3 g/l has been defined as freshwater, based on a study of the relationships between salinity and natural vegetation species richness [7]. Above the salinity threshold of 3 g/l, the species richness decreases dramatically. The brackish water has been defined as water with a salinity ranging from 3 to 15g/l and water with salinity above 15 g/l is considered saline.

The salinity data show that the size of the freshwater lenses changes throughout the year (Figure 13). Below the coastal pine forest and close to sea, the aquifer is salty and there is a 0.5 m thick fresh-brackish water layer at the top of the aquifer, which disappears in summer. Salinity data indicate a counter-intuitive trend underneath the irrigated agricultural land at 2 km from the sea: in summer the salinity is lower than in the winter. This may be explained by the high irrigation rates during the growing season or by infiltration from the irrigation ditch. Inside the historical *Classe* pine forest, at 4 km from the coast, the groundwater is mostly brackish (10 g/l) with a 1 m thick freshwater layer at the top of the aquifer that, in the coastal pinewood, vanishes rapidly. Fresh groundwater lenses within the quarry belt at 7 km from sea have a larger thickness (2 m) but at the bottom of the aquifer saltwater is also present. At a distance of 10 km from the coast, in agricultural land, there is fresh water throughout the whole thickness of the aquifer (Figure 13).

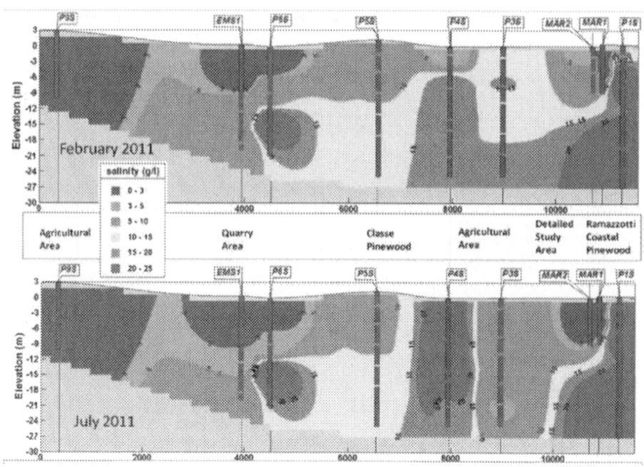

Figure 13. Transect 10 km long from the Adriatic coast on the right towards the mountains on the left. Warm colors represent brackish-saline water whereas cold colors are for freshwater.

On the basis of the cations and anions analyses, alkalinity, and salinity, a groundwater type classification used by Stuyfzand [45] was obtained [11]. Seawater is typically of the S4NaClo type meaning a saline water with High Alkalinity, Na being the major cation, Cl the major anion and a zero BEX indicating a stable water without cation exchange taking place (for data see [11]).

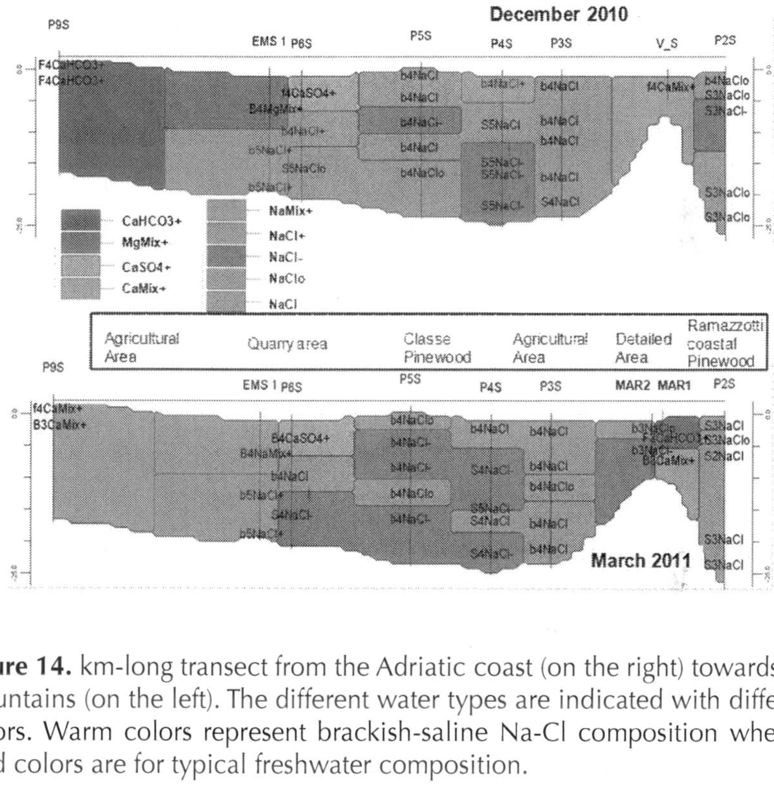

Figure 14. km-long transect from the Adriatic coast (on the right) towards the mountains (on the left). The different water types are indicated with different colors. Warm colors represent brackish-saline Na-Cl composition whereas cold colors are for typical freshwater composition.

Apennine river water is typically of the type (g-F)3CaHCO$_3$ and (f-F)3CaMix, or in other words: fresh water, medium alkalinity and rich in Calcium and varying anions. Po river water coming through the CER is of the type g2CaHCO$_3$ and slightly less alkaline than Apennine river water [11]. The groundwater water type is diagnostic to evaluate if the river-irrigation water has replaced the saline groundwater. Figure 14 shows that the sodium-chloride water type is dominant in the whole section. It is important to note that there are differences between the winter and spring situations in the area studied in detail. This was caused by the addition of new wells in the second monitoring campaign (MAR1 and MAR2 were monitored in April 2011 only). Excluding this effect, two water type changes were observed in the western part of the transect: from calcium-carbonate in December 2010 to calcium-mix in April 2011 in well P9S and from magnesium to sodium in well EMS. The April water type for the area near the irrigation ditch shows, in MAR1 well, a calcium carbonate water type, which is the same chemical composition of river water. Otherwise, in well MAR2 despite its low salinity (Figure 13), the dominant water type is sodium-chloride (Figure 14).

The BEX index for dolomite-rich aquifers as reported by Stuyfzand [45] is used to study whether the groundwater is becoming more or less saline. The BEX is indicated in Figure 15 with a +, - or 0 character after the water type code. The groundwater samples at the top of the aquifer and in winter show a positive BEX index, which indicates freshening (Figure 14).

Rainfall from January to September 2011 recorded in the five pluviometers along the section has a fairly homogeneous value around 400 mm. Total irrigation of a tomato field in the area studied in detail was 600 mm, applied with a rate of 120 mm per month (April-August). The infiltration rate of water into the aquifer, measured from drain gauges located in the tomato field, was 8 mm/m^2 in May, 2 mm/m^2 in June and 0 in the other months (Table 1).

Table 1. Precipitations, irrigation and infiltration measured in tomato parcel from January to September 2011

2011 (unit mm)	**Months**	January	February	March	April	May	June	July	August	September
	Rainfall	33	38	96	15	31	40	43	0	27
	Irrgation	0	0	0	150	120	150	180	90	0
	Infiltration	n.a.	n.a.	n.a.	8	2	0	0	0	0

Water table measurements in April 2011 show that the groundwater is at 0.2 m below sea level in the*Ramazzotti* pinewood and, moving landward, the depth linearly increases up to 1.2 m below sea level (Figure 15). The water level decreases rapidly everywhere during the summer, except for the area surrounding the irrigation channel (Figure 15) where an inverse gradient, with respect to the normal situation, is present during the whole irrigation time (from May to August 2011). As in winter the water level in the irrigation ditch is equal to the groundwater level, during the growing season the water level in the ditch is higher than the groundwater level, creating a hydraulic gradient in the phreatic aquifer that is directed to sea. At the end of the irrigation season (September for the year 2011), the inputs of freshwater end and the water level in the ditch returns to match the groundwater level (Figure 15).

The February 2011 survey results suggest that below the eastern parcel, where the aquifer is phreatic, a 4-m-thick freshwater lens is present before the irrigation season; freshwater is also present west of the ditch but its thickness is only 1,5 m. During both field campaigns in February and July 2011, VES profiles and the depth of the water table and the salinity of the groundwater at the water table were measured by drilling auger holes. The measurements made in February (Figure 16) show that the water level in the ditch is the same of the groundwater and that the whole system has a small gradient towards the pumping station. The freshwater lens vanishes below the *Ramazzotti* pinewood (E8 location) where the the water is salty up to the surface (Figure 16).

During the irrigation period (July 2011), the ditch is full of freshwater. The irrigation of tomatoes in the eastern parcel causes the whole soil profile to be moist. West of the ditch, where the Alfa Alfa was grown, the water table depth increases and also the bottom of the dry soil line becomes deeper (Figure 16). Comparing the VES profilqes made in February 2011 with those made in July 2011 confirms the existence of a well-developed freshwater lens in the eastern part that grows in size during the irrigation season (Figure 16). The largest growth in thickness of the freshwater lens occurs under the ditch and in the western part where the semi-confined aquifer is present.

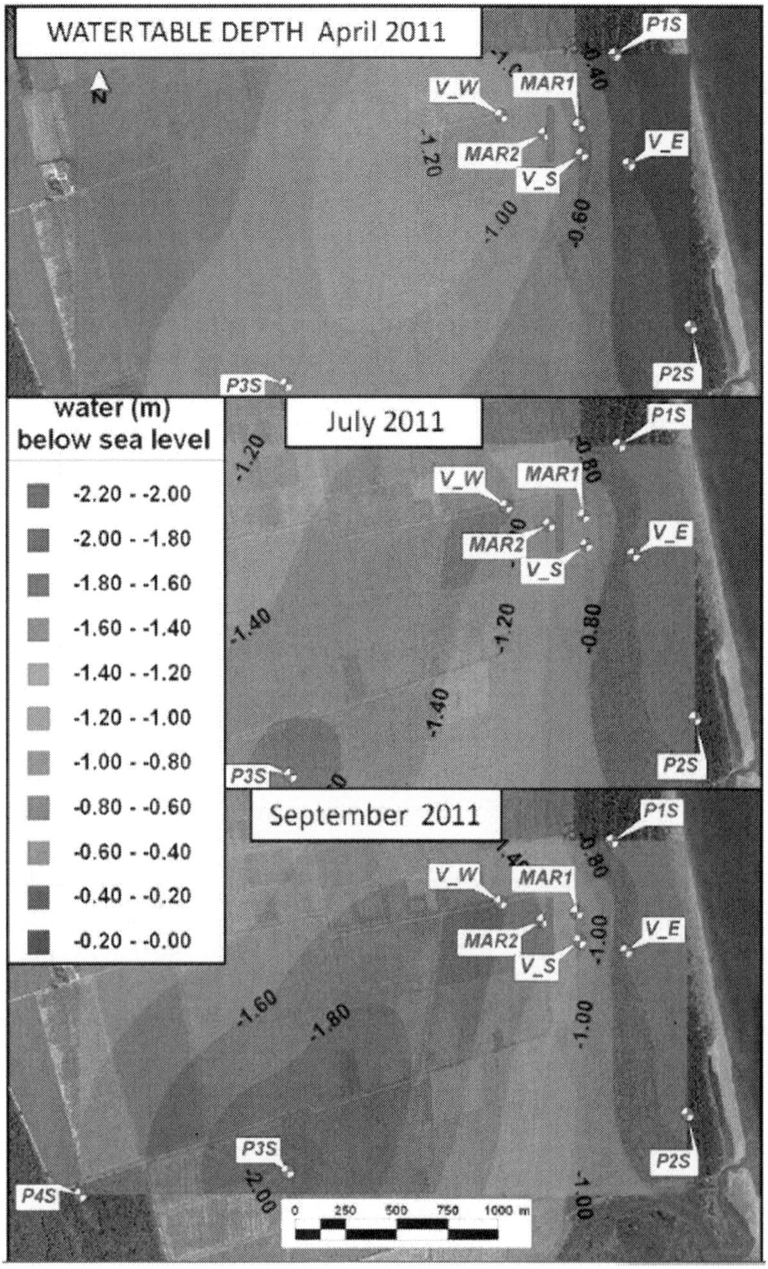

Figure 15. Study area water table map for April, July, and September showing also the irrigation ditch (red line).

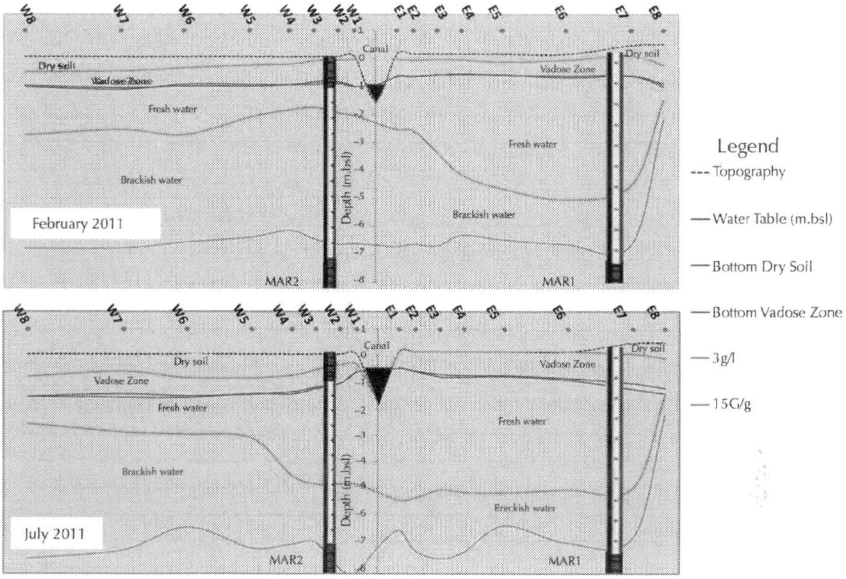

Figure 16. Seasonal comparison of VES results for the agricultural land close to the freshwater storage ditch.

DISCUSSION

As recently reported by IPCC [22] a reduction of freshwater availability and an increase of drought periods length is likely to occur in the Ravenna area during the next 100 years. An increase in temperature with a simultaneous increase of evaporation will cause soil salinization and loss of production. Near Ravenna more than 68 square kilometers are at risk for soil salinization caused by salt water intrusion in the coastal aquifer and along main rivers and channel. A stable isotope analysis shows that the ground and surface water near the city of Ravenna is a mix of sea, rain, and river water. The sea-water is, in part, inherited from the Holocene deepest marine ingression [11]. The aquifer has the shape of a sandy wedge enclosed in clay and this inhibits an efficient flushing of the saline water that is in part left at the bottom of the aquifer from the optimum climate period. Deeper aquifers in the area receive underground recharge from water with a mountain or rainfall sources. Where present, freshwater is located in the recent and paleo-dune belts, because of their high infiltration potential [3],[22],[47]. Where channels and rivers flow on a sandy top layer, there is infiltration of freshwater into the aquifer as along the *Bevano* and *Fiumi Uniti* rivers as well as in the western part of the San Vitale pinewood [11]. There is a small area along the *Bevano* River, where the freshwater lens reaches the bottom of the aquifer but in all other cases the freshwater lenses are floating on top of brackish groundwater. These freshwater lenses are thin (1m maximum), ephemeral (present only during winter) and strongly linked to the climate conditions [3], [22], [48], [49], [50].

Aquifer salinization in our study area is due to the combination of several factors such as low topography, natural and anthropogenic subsidence, mechanical drainage, destruction of the coastal dune belt, and scarcity of freshwater infiltration [11][43][51][52]). As shown in Figure 8, in the whole aquifer, the groundwater head is below sea level in winter also. This situation causes a sea-water encroachment promoted also by the high discharge water volumes of the pumping stations. Based on number of activity hours, we roughly calculated that in the *Quinto* Basin the volume of drainage water was 8 times larger than that of the total rainfall in same basin for the period.

In this contest, with low freshwater availability and poor natural recharge, it is important to look for some technique that is able to counteract soil and groundwater salinization and at the same time encourages water management practices that prevent nitrate leaching, safeguard the economics of agriculture, and save the freshwater resources.

Our attention was focused on a small agricultural area at a distance of 500m from the shoreline, where despite all previous studies [22] [51][52][53], the presence of a freshwater lens throughout the year, was found. Although this area is affected by drainage, low topography, and subsidence as the rest of the Ravenna province, the presence of a sandy top layer, the horticulture with irrigation, and a ditch used for freshwater storage, all promoted aquifer recharge. The chemical analysis and electrical surveys showed that the size of the freshwater lens is larger during the summer period (Figure 13,Figure 14 and Figure 16).

In many other countries worldwide, managed aquifer recharge (MAR) is used to store fresh river or rain water in the underground of riverbanks or dunes for later use as drinking water [54].

Managed Aquifer Recharge or MAR is also used for re-pressurization of aquifers subject to falling water levels, declining yields, saline intrusion or land subsidence. Examples of these from around the world include different technologies such as aquifer storage and recovery (ASR), aquifer storage transfer and recovery (ASTR), bank and dune filtration, infiltration ponds, percolation tanks, rainwater harvesting, and soil-aquifer treatment (SAT) [55]-[61]. All above mentioned technologies have a cost that sometimes may be not sustainable by the local economy. In countries such as India, China, and Pakistan artificial recharge from ditches is affordable for the cultivation of rice, but this is usually done far from the coast ([61][62][63][64]). As reported by Gale 2000 ([61]), in many countries, the seepage from ditches or channels is considered a loss of water and they are studying systems to solve this problem. Irrigation ditches, on the other hand, may also play an ecological role: some authors report that ditches support the biodiversity and associated ecosystem services ([65]) and that they promote retention and removal of agrochemicals, decrease soil erosion, and allow the early stage of development for fish, amphibians, and reptiles [66]. Another aspect to consider is the low cost for their construction and management (removal of excess vegetation to restore efficient drainage) [66]. In particular for our area a dense network of ditches is already present and their building costs could be very low or absent.

Irrigation practices tend to become more and more efficient by letting go less and less water into soil-evaporation. The driving forces for that are the scarcity of freshwater and economic incentives. One might say that the water left in the ditch to recharge the aquifer is water that is 'lost'. One of the main challenges in water resources and agricultural management in coastal Mediterranean zones will be to define priorities that may vary with the particular area. In one place it may be important to use as little water as possible for irrigation not leaving anything to waste or end up in the aquifer underneath the top soil. In other places instead, especially in the coastal zone, irrigation water may be left to seep from the ditches into the surface aquifer to create freshwater lenses that sustain the agricultural land as well as the natural areas and at the same time counteract seawater intrusion.

The high salinity of the coastal aquifer near Ravenna is in part inherited from its geological history. Therefore it is unlikely that the aquifer will be ever used as a source for drinking water. By infiltrating (excess) river water on purpose through ditches as the one studied in this paper, however, freshwater lenses can be created that may prevent soil salinization and serve agricultural crops as well as natural areas. Our studied ditch has the same dimension as the drainage channels in the area. The only difference is that it is a closed basin used for storage irrigation water during the summer. By using the existing hydraulic infrastructure for managed aquifer recharge in ditches, more freshwater would flow into the aquifer instead of being drained away by the pumping machines. This system of aquifer artificial recharge would be viable in similar coastal settings throughout the world.

CONCLUSIONS

Despite the fact that there are many technological, agricultural, economic, and water availability factors that are dependent on the local situation and farming practices (crop rotations, volume and timing of irrigation, tillage, economic forcing, etc.) and that the length of the study was of only one year, so that no long term trends can be captured by the analysis presented, some interesting observations have been made:

- The land use in the Ravenna coastal area is complex; natural areas (pine forests, lagoons and wetlands) and industrial activities are placed in a landscape dominated by agriculture.
- The Low topography, the high rates of subsidence, the salty phreatic aquifer require land drainage to allow for any activity. The result is that drainage allows for agriculture but it decreases the recharge of the aquifer, removes freshwater from the top of the aquifer and increases up-welling of saline water from below.
- The surface hydrology is controlled by a high density network of channels used mainly for drainage but locally for irrigation.
- Due to different geological events, different kinds of sediments were deposited in the area; therefore special soils were developed and crops distribution as well as irrigation water requirements depends strictly from the soil characteristics.
- Irrigation is essential for horticultural crops, which are mainly located on sandy soil where the connection with the aquifer below is guaranteed.

- The water sprayed on the land for irrigation does not infiltrate into the aquifer and this is a good practice to avoid nitrate leaching and contamination of the groundwater. In the area studied in detail, the only irrigation water added to the aquifer is from seepage from the storage ditch.
- The case study shows that the presence of a ditch, full of freshwater for 5 months during the year, generates a permanent freshwater bubble 4 m in thickness, at 500 m distance from the sea. This shows that combined irrigation and drainage management, considering the dense network of existing channels and different types of soil, can be of benefit for the agriculture for soil protection, storage of freshwater resources, and to counteract sea-water intrusion.

ACKNOWLEDGEMENT

We thank the farmer of the "Cooperativa Braccianti di Campiano" for their helpfulness and kindness; we thank also University of Bologna, the Autorità dei Bacini Romagnoli and the Municipality of Ravenna for funding the several projects that form the basis of this chapter. Special thanks to Murugan Ramasamy and Sathish Sadhasivam for their fundamental support in field work.

REFERENCES

1. G. Toth, L. Montanarella, E. Rusco, 2008Updated Map Of Salt Affected Soils In The European Union. Threats To Soil Quality In Europe, Office For Official Publications Of The European Communities, Luxembourg (2008), 6174
2. P. G. B. Louw, S. De Eeman, B. Siemon, B. R. Voortman, J. Gunnink, E. S. Baaren, G. H. P. Van And, Essink. Oude, 2011Shallow Rainwater Lenses In Deltaic Areas With Saline Seepage, Hydrol. Earth Syst. Sci. Discuss., 876577707
3. M. Antonellini, P. Mollema, B. Giambastiani, E. Banzola, K. Bishop, L. Caruso, A. Minchio, L. Pellegrini, M. Sabia, E. Ulazzi, Gabbianelli. G. And, 2008Salt Water Intrusion In The Coastal Aquifer Of The Southern Po-Plain, Italy. Hydrogeology Journal 1615411556DoiS10040-008-0319-9.

4. E. Custodio, 2010Coastal Aquifers Of Europe: An Overview. Hydrogeology Journal, 181269280DOI:S10040-009-0496-1
5. FAO2002Agricultural Drainage Water Management In Arid And Semi-Arid Areas Ftp://Ftp.Fao.Org/Docrep/Fao/005/Y4263e/Y4263e11.Pdf
6. E. V. Maas, G. J. Hoffman, 1977Crop Salt Tolerance, Current Assessment J. Irrig. Drain. Div., 103, 115134
7. M. Antonellini, Mollema. P. And, 2010Impact Of Groundwater Salinity On Vegetation Species Richness In The Coastal Pine Forests And Wetlands Of Ravenna, Italy. Ecological Engineering. 236 (9), 1201-1211, Doi:10.1016/J.Ecoleng.2009.12.007.
8. I. Hutchinson, 1991Salinity Tolerances Of Plants Of Estuarine Wetlands And Associated Uplands. Washington State Department Of Ecology Report, Washington, USA.
9. IPWEA,2001The Local Government Salinity Management Handbook. Institute Of Public Works Engineering Australia (IPWEA), Australia.
10. T. Lombardi, B. Lupi, 2006Effect Of Salinity On The Germination And Growth Of Hordeum Secalinum Schreber (Poaceae) In Relation To The Seeds After Ripening Time. Atti Della Società Toscana Di Scienze Naturali, Memorie, Serie B 1133742
11. P. Mollema, A. Antonellini, E. Dinelli, G. Gabbianelli, N. Greggio, P. J. Stuyfzand, 2012Identification Of Fresh And Salt Water Intrusion Types In The Upper Ravenna Coastal Aquifer (Italy) Via Hydrochemical Systems Analysis. Submitted to Applied Geochemistry.
12. Istat2000Censimento Generale Dell'agricoltura http://www.census.istat.it
13. Provincia Di Ravenna2006Piano Territoriale Di Coordinamento Provinciale
14. V. Marconi, M. Antonellini, E. Balugani, E. Dinelli, 2011Hydrogeochemical Characterization Of Small Coastal Wetlands And Forests In The Southern Po Plain (Northern Italy). Ecohydrol. 4
15. P. Mollema, M. Antonellini, G. Gabbianelli, M. Laghi, V. Marconi, A. Minchio, 2012Climate And Water Budget Change Of A

Mediterranean Coastal Watershed, Ravenna, Italy. Environmental Earth Sciences. 65257276Doi:S12665-011-1088-7..

16. Fao.1990Guidelines For Soil Profile Description. Third Edition (Revised). Soil Resources, Management And Conservation Service, Land And Water Development Division, Fao, Rome.
17. Soil Survey Staff.1996Keys To Soil Taxonomy. Seventh Edition. United States Department Of Agriculture, Washington D.C.
18. D. A. Rose, F. Konukcu, J. W. Gowing, 2005Effect Of Water table Depth On Evaporation And Salt Accumulation From Saline Groundwater. Aust. J. Soil Res. 43565573
19. Andrej. T. Il'ichev, Tsypkin. And, G. George, Pritchard. And, And. David, Chris. N. Richardson, 2008Instability Of The Salinity Profile During The Evaporation Of Saline Groundwater. Journal Of Fluid Mechanics, 614. 871040022-1120
20. N. Filippi, 1994Carta Pedologica (Soil Map) 1: 250.000. Stampata Dalla Regione Emilia Romagna, Italy
21. S. Corticelli, 2003Carta Del Uso Del Suolo (Land Use Map). Copertura Vettoriale. Stampata Dalla Regione Emilia Romagna, Italy
22. B. M. S. Giambastiani, M. Antonellini, Essink. G. H. P. Oude, R. J. Stuurman, 2007Salt Water Intrusion In The Unconfined Coastal Aquifer Of Ravenna (Italy): A Numerical Model. Journal Of Hydrology 34094104
23. B. M. S. Giambastiani, P. N. Mollema, M. Antonellini, 2008Groundwater Management In The Northern Adriatic Coast (Ravenna, Italy): New Strategies To Protect The Coastal Aquifer From Saltwater Intrusion. In: Groundwater: Modelling, Management And Contamination. Authors: König L.F. And Weiss J.L.; Nova Publishers
24. Consorzio Di Bonifica Della Romagna2002Il Comprensorio E Le Infrastrutture http://www.bonificaromagna.it/Cartografia.ìt/html
25. Regione Emilia Romagna, Agip,2009Riserve Idriche Sotterranee Della Regione Emilia Romagna, Technical Report. Regione Emilia Romagna, Firenze
26. F. Giorgi, P. Lionello, 2008Climate Change Projections For The Mediterranean Region. Global Planet Change 6390104

27. G. Antolini, V. Marletto, F. Tomei, V. Pavan, R. Tomozeiu, 2008Atlante Idroclimatico Dell'emilia-Romagna 19612008

28. A. Magnan, B. Garnaud, R. Billé, F. Gemenne, Hallegatte. S. Iddri, Cired-Météo, 2009The Future Of The Mediterrenean. From Impacts Of Climate Change To Adaptation Issues. Institut Du Développement Durable Et Des Relations Internationals, Paris, France, 43 Pp.

29. A. Amorosi, M. L. Colalongo, G. Pasini, D. Preti, 1999Sedimentary Response To Late Quaternary Sea-Level Changes In The Romagna Coastal Plain (Northern Italy). Sedimentology, 4699121

30. R. Catalano, E. Di Stefano, S. Infuso, A. Sulli, P. R. Vail, F. P. Vitale, 1995Sequence And Systems Tracts Calibrations On High Resolution Bio-Chronostratigraphic Scheme: The Central Mediterranean Plio-Pleistocene Record. In: Mesozoic-Cenozoic Sequence Stratigraphy Of Western European Basins (Ed. By P.C. De Graciansky Et Al.). Spec. Publ. Soc. Econ. Paleont. Miner

31. A. Rizzini, 1974Holocene Sedimentary Cycle And Heavy Mineral Distribution, Romagna-Marche Coastal Plain, Italy: Sedimentary Geology, 111737

32. L. M. J. U. Van Straaten, 1970Holocene And Late Pleistocene Sedimentation In The Adriatic Sea. Geo/. Runsch. 60106131

33. Giambastiani B.M.S.2003Geomorfologia Del Litorale Ravennate: I Sistemi Dunosi. Unpublished Msc Thesis: University Of Bologna

34. F. Stecchi, A. Minchio, C. Del Grande, Gabbianelli. G. And, 2003Historical And Recent Evolution Of The Reno Rover Mouth And Adjacent Areas. In. Gabbianelli G. And Sangiorgi F. (Editors). 4Of Proceedings Of The Sixth International Conference On The Medditeranean Coastal Environment, Medcoast 03, 7-11 October 2003, Ravenna Italy, 5567

35. M. Bondesan, Gb. Castiglioni, C. Elmi, G. Gabbianelli, R. Marocco, Pa. Pirazzoli, A. Tomasin, 1995b Coastal Areas At Risk From Storm Surges And Sea Level Rise In North-Eastern Italy. J Coast Res 1113541379

36. A. Amorosi, M. L. Colalongo, F. Fiorini, F. Fusco, G. Pasini, S. C. Vaiani, Sarti, G.,2004Palaeogeographic And Paleoclimatic Evolution Of The Po Plain From 150-Ky Core Records. Global And Planetary Change, 405578

37. A. Correggiari, M. Roveri, F. Trincardi, 1996Late-Pleistocene And Holocene Evolution Of The North Adriatic Sea. Late-Glacial And Early Holocene Climatic And Environmental Changes In Italy: Il Quaternario: Italian Journal Of Quaternary Sciences 9697704
38. A. Amorosi, M. L. Colalongo, F. Fiorini, F. Fusco, G. Pasini, S. C. Vaiani, Sarti, G.,2004Palaeogeographic And Paleoclimatic Evolution Of The Po Plain From 150-Ky Core Records. Global And Planetary Change, 405578
39. L. Marchesini, A. Amorosi, U. Cibin, G. G. Zuffa, E. Spadafora, D. Preti, 2000Sand Composition And Sedimentary Evolution Of A Late Quaternary Depositional Sequence, Northwestern Adriatic Coast, Italy. Journal Of Sedimentary Research 70 (4), 829-838.
40. D. K. Bulter, 2005Near-Surface Geophysics, (Seg Investigations In Geophysics Series; N°13), Library Of Congress; Pag. 46
41. H. Burger, Robert, 1992Exploration Geophysics Of The Shallow Subsurface. Prentice Hall Pagg. 291295
42. P. Celico, 1998Prospezioni Idrogelogiche. Vol I, Liguori Editore, V° Ristampa 1998 Pag 243
43. D. Capo, 2011Phd Thesis "Caratterizzazione Della Falda Freatica Costiera Ravennate" In Press
44. M. H. Loke, 2001Electrical Imaging Surveys For Environmental And Engineering Studies- A Practical Guide To 2D And 3-D Surveys. Residinv Manual, Iris Instruments; 67 Pp.
45. P. J. Stuyfzand, 1989A New Hydrochemical Classification Of Watertypes. Iahs Publ. 1828998
46. IPCC Technical Paper Vi.2008Climate Change And Water-Intergovernmental Panel On Climate Change http://www.ipcc.ch/pdf/Technical-Papers/Climate-Change-Water-En.Pdf
47. V. Marconi, M. Antonellini, E. Balugani, E. Dinelli, 2011Hydrogeochemical Characterization Of Small Coastal Wetlands And Forests In The Southern Po Plain (Northern Italy). Ecohydrol. 4
48. B. Capaccioni, M. Didero, C. Paletta, L. Didero, 2005Saline Intrusion And Refreshening In A Multilayer Coastal Aquifer In The Catania Plain (Sicily, Southern Italy): Dynamics Of Degradation Processes According To The Hydrochemical Characteristics Of Groundwater. Journal of Hydrol 307116

49. Essink. G. H. P. Oude, 2001Improving Fresh Groundwater Supply-Problems And Solutions: Ocean And Coastal Management 44429449
50. Pp. Chot, Sc. Dekker, A. Poot, 2004The Dynamic Form Of Rainwater Lenses In Drained Fens: Journal Of Hydrology 2937484
51. V. Marconi, M. Antonellini, M. Laghi, A. Minchio, D. Savelli, 2008A Water Table Fluctuations Model In Sandy Soil Below A Coastal Pine Forest. Geophysical Researches Abstract 10: Egu201005761
52. Marconi V., PhD Thesis2010Hydrogeochemical Characterization Of Small Coastal Wetlands And Forests In The Southern Po Plain (Northern Italy).
53. V. Marconi, 2011Effetti Della Vegetazione E Del Drenaggio, Sull'intrusione Salina Nell'acquifero Freatico Costiero Della Zona Compresa Fra Foce Dei Fiumi Uniti E Foce Bevano (Ravenna). Dottorato Di Ricerca In Scienze Ambientali: Tutela E Gestione Delle Risorse Naturali, Ciclo XXIII. Università Di Bologna
54. P. J. Stuyfzand, 1999Patterns In Groundwater Chemistry Resulting From Groundwater Flow. Hydrogeology Journal 7 1527
55. Böhlke Jk2002Groundwater Recharge And Agricultural Contamination Hydrogeology Journal (2002) 10:153-179 DoiS10040-001-0183-3
56. J. Boonstra, M. N. Bhutta, 1996Groundwater Recharge In Irrigated Agriculture: The Theory And Practice Of Inverse Modeling Journal Of Hydrology 174 357 374
57. H. Bouwer, 199, 1997Role Of Groundwater Recharge And Water Reuse In Integrated Water Management. Arabian J. Sci. Eng., 22123131
58. H. Bouwer, 1999Chapter 24: Artificial Recharge Of Groundwater: Systems, Design, And Management. Hydraulic Design Handbook, L. W. Mays, Ed., Mcgraw-Hill, New York, 24.124
59. H. Bouwer, P. Fox, P. Westerhoff, Drewes. J. E. And, 1999Integrating Water Management And Reuse: Causes For Concern? Water. Qual. Internat., 19991922
60. P. Dillon, 2005Future Management Of Aquifer Recharge Hydrogeol J. 13313316

61. I. Gale, 2000Strategies For Managed Aquifer Recharge (Mar) In Semi-Arid Areas International Association Of Hydrogeologists Commission On Management Of Aquifer Recharge Iah- Mar International Hydrological Programme (Ihp) Unesco Division Of Water Sciences www.iah.org/recharge

62. K. S. Meijer, 2000Impacts Of Concrete Lining Of Irrigation Canals, Uda Walawe, Sri Lanka. M.Sc. Thesis, University Of Twente, The Netherlands.

63. K. Meijer, E. Boelee, D. Augustijn, I. Van Der Molen, 2006Impacts Of Concrete Lining Of Irrigation Canals On Availability Of Water For Domestic Use In Southern Sri Lanka. Agric. Water Manage. 83243251

64. Liqiang Yao, Shaoyuan Feng, Xiaomin Mao, Zailin Huo, Shaozhong Kang, D.A. Barry.2012Coupled Effects Of Canal Lining And Multi-Layered Soil Structure On Canal Seepage And Soil Water Dynamics Journal Of Hydrology 430-431, 91-102

65. R. B. Bradbury, W. B. Kirby, 2006Farmland Birds And Resource Protection In The Uk: Cross-Cutting Solutions For Multi-Functional Farming? Biological Conservation, 129, 530542

66. I. Herzon, J. Helenius, 2008Agricultural Drainage Ditches, Their Biological Importance And Functioning Biological Conservation 14111711183

Chapter 4

Dryland Soil Hydrological Processes and Their Impacts on the Nitrogen Balance in a Soil-Maize System of a Freeze-Thawing Agricultural Area

Wei Ouyang[1*], Siyang Chen[2], Guanqing Cai[1], Fanghua Hao[1]

[1]School of Environment, State Key Laboratory of Water Environment Simulation, Beijing Normal University, Beijing, China,

[2]Marine Monitoring and Forecasting Center of Zhejiang, Hangzhou, China

ABSTRACT

Understanding the fates of soil hydrological processes and nitrogen (N) is essential for optimizing the water and N in a dryland crop system with the goal of obtaining a maximum yield. Few investigations have addressed the dynamics of dryland N and its association with the soil hydrological process in a freeze-thawing agricultural area. With the daily monitoring of soil water content and acquisition rates at 15, 30, 60 and 90 cm depths, the soil hydrological process with the influence of rainfall was identified. The temporal-vertical soil water storage analysis indicated the local *albic* soil texture provided a stable soil water condition for maize growth with the rainfall as the only water source. Soil storage water averages at 0–20, 20–40 and 40–60 cm were observed to be 490.2, 593.8, and 358 m^3 ha^{-1}, respectively, during the growing season. The evapo-transpiration (ET), rainfall, and water loss analysis demonstrated that these factors increased in same temporal pattern and provided necessary water conditions for maize growth in a short period. The dry weight and N concentration of maize organs (root, leaf, stem, tassel, and grain) demonstrated the N accumulation increased to a peak in the maturity period and that grain had the most N. The maximum N accumulative rate reached about 500 mg $m^{-2}d^{-1}$ in leaves and grain. Over the entire growing season, the soil nitrate N decreased by amounts ranging from 48.9 kg N ha^{-1} to 65.3 kg N ha^{-1} over the 90 cm profile and the loss of ammonia-N ranged from 9.79 to 12.69 kg N ha^{-1}. With soil water loss and N balance calculation, the N usage efficiency (*NUE*) over the 0–90 cm soil profile was 43%. The soil hydrological process due to special *soil* texture and the temporal features of rainfall determined the maize growth in the freeze-thawing agricultural area.

INTRODUCTION

During the agricultural tillage management, the maximum crop production and minimum diffuse nitrogen (N) loading are the priority issues that need to be considered at the same time [1],[2]. It is essential to understand the fates of soil water (SW) and N in agricultural systems in order to attain higher crop yields and N usage efficiency [3]. In developing countries, there is much political and commercial pressure for controlling the N application with the sacrifice of the crop harvest [4]. The dryland agriculture in the Sanjian Plain, Northeast China, is a key food base and the most water limited agricultural zone in China [5]. In this freeze-thawing area, the soil N and water efficiency are the keys for dryland tillage sustainability due to the short growing season. However, there are few reports about the dryland soil hydrological process and N use efficiency in freeze-thawing agricultural areas [6].

The soil nutrient N is a major limiting factor for crop growth and production, which is directly related to the N fertilisation and soil N background level [7]. The soil microbial community, which affects the soil N cycle [8], is less active in this freeze-thawing area. The application of chemical N is a basic agricultural practice in worldwide, which causes excessive discharge of N to the aquatic environment [9]. The European Union agricultural landscape contributes about 55% of the diffuse pollution for water eutrophication. This is a major emerging environmental issue in the developing counties [10]. The soil eco-hydrological process is an essential channel for N transport in the soil-crop system.

Plant organs have different allocation rates with N during growth, and the canopy green area has higher N absorption than the other organs during the jointing period [11]. The N accumulations in the upper leaves of the canopy also follow the eco-hydrological pattern in farmland [12]. Some models have been developed to simulate the N absorption in different organs during crop growth with the impacts of the solar radiation, rainfall, temperature, soil hydrology and N concentration [13]. With the temporal pattern of N accumulation in crop organs, the N cycle and efficiency can also be analyzed. The diffuse N discharge from agricultural systems depends on diverse factors, including the climatic, soil properties and agronomic features [14],[15]. The tillage practice, soil hydrological process and N usage efficiency are the main potential factors to consider when the goal is to reduce the agricultural diffuse N pollution [16]. The dryland maize farmland in a freeze-thawing agricultural area presents different N discharge patterns due to the special hydrological process [17].

The soil hydrological process is the bridge for diffuse N loss from soil to water bodies [18]. The nitrate-N is the dominant type of diffuse N in the soil profile and it can move swiftly downstream flow with the rainfall through the soil pathways [19]. The soil hydrological process is also closely related to the soil texture, which is the medium for water movement and crop growth. In the freeze-thawing dryland, the soil has short-term, transitional phase and rainfall is the only water source for N movement with crop growth [20]. The soil pattern in the study area is meadow*albic* soil (*Albic Luvisols*) and has a 20–30 cm depth impermeable stratum. The *albic* soil provides a stable soil water (SW) condition in the crop root zone and interfaces with the soil hydrological process. With the consideration of high N efficiency, it is necessary to analyze the special pattern of soil hydrology and its impact on the N efficiency [21]. Evaluating the response of N loss to the combination of rainfall and the soil hydrological process can help to identify the interactions of water with N and reduce diffuse N pollution.

The general objective is to maximize the N efficiency in tillage practices. Typically, of about 20%–70% of N is lost from soil-crop systems and this loss can cause other environmental issues [22]. The quality of soil properties will decline under the combination of intensive application of N fertilizer and low N efficiency; this in turn, will also deliver some risk to water [23]. There are several methods to improve the N efficiency, including accurate fertilizer application and irrigation [24],[25]. The dryland soil hydrological process in a freeze-thawing agricultural area presents a special pattern and also affects the N efficiency [26]. This study aims (1) to identify the temporal-vertical dynamics of dryland SW within an *albic* soil layer; (2) to express the dry weight and N accumulation of maize organs in dryland of a freeze-thawing agricultural area; (3) to explore the variation of soil loss, potential N loss, and N use efficiency (*NUE*) in dryland during the growing season.

MATERIALS AND METHODS

Ethics Statement

The field experiment was conducted in the farmland of Bawujiu Farm, which has been rented by Wang Dongli for 50 years. He is also the contact person for future permission. No specific permissions were required for these locations. The field studies did not involve endangered or protected species. This study did not involve vertebrate. The specific location of our study is 133° 50′–134° 33′ E, 47° 18′47°50′ N.

Study Area Description

The case study area was conducted on a farm in the north-eastern part of China. The farm has a long history and on the east is adjoined to Russia

(Fig. 1). The observation site is located in the southern part of the Farm and the elevation is 48 m. The only water source for dryland tillage is precipitation and maize is the dominant crop. This area the temperate continental monsoon climate has an average annual precipitation of 588 mm and an average yearly temperature of 2.94 °C [27]. From October to the following April, the temperature is below zero and, as a result, the soil freeze-thawing process affects the tillage process. The crop growing season is from May to October.

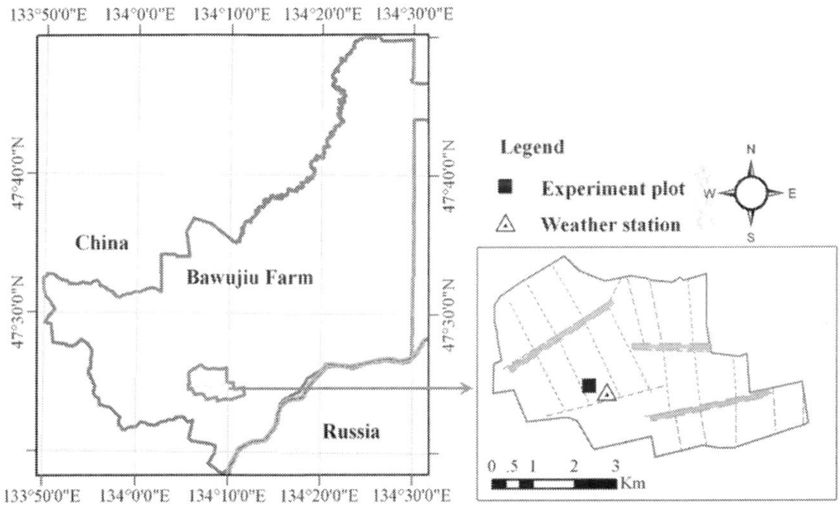

Figure 1: Location of the experimental dryland site in Northeast of China.

Field Monitoring and Soil Sample Collection

In order to identify soil physical-chemical properties, soil samples at four depths (0–15, 15–30, 30–60 and 60–90 cm) were collected in 2010. The samples from three points chosen at random in each layer were mixed into a composite soil sample and placed in plastic bags for lab measurements. The soil samples were collected before sowing and after harvest. Soil particle size distributions were measured by laser diffraction after the removal of organic residue (MasterSizer S, Malvern Instruments, Malvern, UK). The soil's total nitrogen (TN) and organic carbon (OC) concentrations were measured with a CHN Elemental Analyser (Euro VectorS.P.A EA3000, 136 Milan, Italy) [28]. The ammonia N in rainfall was analyzed with the aid of Nessler's reagent and the nitrate-N concentration was determined by the Ultraviolet Spectro Photometric Method. The soil's pH was determined with a pH meter (METTLER TOLEDO, Switzerland) after mixing the fresh soil with deionised water (1:2.5, w/v) [24]. The soil N stock at four depths was calculated with the following equation,

$$Stock_i = BD_i * H_i * C_i * 10^5 \tag{1}$$

where $Stock_i$ is the soil N storage at layer i (kg/m^2), BD_i is the soil bulk density at layer i(g/cm^3), H_i is the soil depth at layer i (cm), and C_i is the soil N concentration at layer i (mg/kg).

For the soil hydrological process monitoring, the SW collector head (suction cup: Teflon and quartz; OD 21 mm×L 95 mm; porosity: 2 μm; conductivity: 3.31×10^{-6} mm sec^{-1}, PRENART, Danmark) was set at 15, 30, 60 and 90 cm depths. It was covered with a mixed solution of quartz powder and stirred, which can cause a 0.05 MPa vacuum pressure and in direct contact with the soil. The collector tail was connected with PVC pipe and the water samples were collected in bottles by a portable manual vacuum pump. The soil water storage was calculated with soil volumetric water content at different depths. The mean soil water storage at each layer was defined as the average of SW in the upper and lower soil layers. The evapo-transpiration in dryland is calculated with the Bowen ratio and energy balance (BREB). The SW balance was calculated with the following equation [29],

$$SL = R - I_r - ET - \Delta S \tag{2}$$

where SL is the soil water loss (mm), R is the precipitation (mm), ET is the evapo-transpiration (mm), I_r is the canopy interception (mm), and ΔS is the soil water storage change (mm).

The local meteorological characteristics were monitored with a ZENO station (Coastal, Seattle, WA, USA). Automated soil temperature sensors (Thermistor, Coastal, Seattle, WA, USA) and soil volumetric water content sensors (TDR type, Coastal, Seattle, WA, USA) were placed in each of the four different soil layers. The monitoring sensors were repeated at double at horizontal 0.5 m space.

Crop Sample Collection and Measurement

The local dryland is intensively cultivated with the aid of mechanized equipment and the maize (*Zea mays L.*) planting density is 75,000 plants ha^{-1}. The detailed maize planting date, harvest date, relevant tillage and fertilization practices are listed in Table 1. The maize samples in 2011 were collected monthly and the sample in 2010 were only collected in October due to the limited experimental conditions. The detailed process of sample collection and monitoring of crop features (height (CH), leaf area index (LAI) and root depth (RD)) are listed in a previously published paper [24]. The maize samples were taken to the lab immediately and the root, leaves, stems, tassels, and grains were separated. The samples were washed clean and then dried for 48 h at 65°C. The dry weight of organ samples was determined after fixing the temperature for 30 min at 105°C. The dried maize samples were crushed and sieved for crop organ N content (%) analysis (Vario EL, Elementar Co. Ltd., Germany). The N uptake amount by crop was calculated based on crop dry weight and plant N content (%).

Table 1: Dryland tillage practices and fertilization information

Crop year	Planting date	Harvest date	Tillage prior to planting	Main fertilisation (kg ha^{-1})
2010	8/Jun.	9/Oct.	Tilling	525
2011	30/May.	5/Oct.		525
Detailed fertilisation data and amounts * (kg ha^{-1})				
2010	2011	N	P	K
8/Jun.	30/May.	71.25	58.65	27.23
27/Jun.	20/Jun.	34.50	13.80	10.02
5/Jul.	1/Jul.	29.45		

Nitrogen Supply and Loss Calculation

The N use efficiency (*NUE*) in a crop-soil system is defined as the ratio of N uptake (N_c) by a crop to N supply (N_s) in a system [30]. The N_s was defined as the total available N to the crop, which included fertilizer N (N_f), mineralized N (N_m), initial mineralized N in soil ($N_{min\ initial}$), N fixed by soil (N_x) and N deposition (N_d) from rainfall. The N_s is calculated as:

$$N_s = N_f + N_{\text{min initial}} + N_m + N_x + N_d \quad (3)$$

The field N_m is calculated with the crop-soil N equation [31]:

$$N_m + (N_{\text{min initial}} - N_{\text{min final}}) = (N_l + N_c + N_{gl}) - (N_f + N_d + N_x) \quad (4)$$

where N_m is the mineralized N, $N_{min\ initial}$ is the initial mineralized N concentration in soil, $N_{min\ final}$ is the final mineralized N concentration in soil, N_l is the soil loss N, N_{gl} is the N loss in gas.N_c is the N crop absorption, N_x is the N fixed by soil, and N_d is the N deposition.

The mineralized N is calculated with the following equation,

$$N_m = (N_l + N_c) - (N_f + N_d) - (N_{\text{min initial}} - N_{\text{min final}}) \quad (5)$$

The potential soil water N loss (N_l) is defined as the total loss of nitrate and ammonia, which was estimated with the N concentration in SW and the water volume in dryland [32]. Based on the SW loss and the N concentration in SW at different depths, the soil N loss by water was calculated.

Data Analyses

Data analysis was performed with Sigma plot 10.0 and SPSS 16.0 software. The differences in dry weight and their percentages on crop organs (including root, leaf, stem, tassel, grain) in different growing stages were tested with ANOVA. The difference in the crop N uptake allocation proportion, the N accumulation in different organs, and the mineralized N concentration in different soil layers were statistically analyzed by ANOVA. The multiple comparisons were analyzed with the Duncan method.

RESULTS

Temporal-Spatial Distribution of Soil Water in Dryland

With the soil-maize water monitoring system, the daily precipitation, SW acquisition rate, and SW content in the entire growing period was determined (Fig. 2). The monthly precipitation in the years 2010 and 2011 displayed a similar pattern, except for the lower value in September 2010. The precipitation occurred mainly between July and September, which was also the prime period for maize growth. With the light precipitation in April and May, the maize seed was sowed in moist soil. The precipitation in August and September of 2010 was 64 mm and 120 mm less than in the corresponding months of 2011. The cumulative precipitation from April to September in 2011 was 575 mm, which was 141 mm more than that in 2010. The precipitation was the only water source for the dryland in this freeze-thawing agricultural area, which directly impacted the SW acquisition rate. The SW could be collected only after the rainfall and the vertical difference in the SW acquisition rate was slight (around 0.002 cm3 s−1). In the entire monitoring period, the biggest acquisition rate occurred at the 90 cm depth after the rainfall in early September. The direct reason for it was that only smaller amount of the water can move deeply in the vertical direction after the absorption by roots.

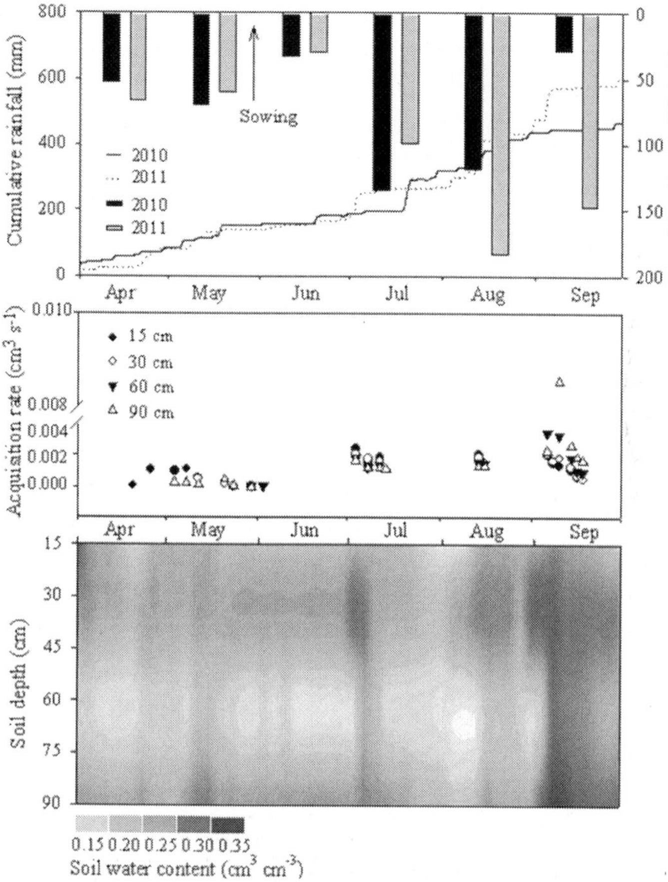

Figure 2: Daily precipitation in 2010 and 2011, and acquisition rate, and soil hydrological pattern of dryland in 2011.

The SW content increased in a stable manner during the growth period and also coincided with the accumulative precipitation. In all of June, the SW content at the 30 cm depth was the highest, the reason for this was that uptake of water in growing roots resulted in water accumulation in the rhizosphere. Due to the 73.6 mm rainfall on July 3, the SW content at the 30 cm depth increased by 20% to 0.30 cm^3 cm^{-3}. In the next two days, the SW content was transported in the vertical direction and it increased to 0.30 cm^3 cm^{-3} in the deep layer. This vertical distribution also confirmed the retention role of roots in the early growing season. The surface SW content increased significantly after August due to the precipitation and then was transported to the deep soil layers. Later, the impact of the crop roots decreased and the soil texture was the main factor for SW transport in September.

In order to further analyse SW content fluctuation in the temporal-vertical dimensions, a statistical analysis was performed on the data from the soil layers during the growing season (30 May–30 September) (Table 2). The SW content averages at the 15 cm and 60 cm depths were very close with values of 0.245 and 0.230 $cm^3\ cm^{-3}$, respectively. The mean SW content at the 30 cm and 90 cm depths were higher than at the other two layers with values of 0.272 $cm^3\ cm^{-3}$ and 0.280 $cm^3\ cm^{-3}$, respectively. The close relationship between the SW acquisition rate and the SW content that was shown in Fig. 2, was also confirmed by the statistical results. The SW acquisition rate had the same vertical pattern as the SW content and the higher values appeared at the 30 and 90 cm depths. Before maize seeding (May 30), the SW could be collected when the SW content reached about 0.25 $cm^3\ cm^{-3}$. After maize seeding, the SW could be collected when it reached a content of over 0.30 $cm^3\ cm^{-3}$. The SW acquisition rate was greatly and primarily influenced by rainfall.

Table 2: Statistical analysis of soil water content and acquisition rate at four depths during maize growing period

Depth (cm)								
Index	Soil water content ($cm^3\ cm^{-3}$)				Soil water acquisition rate ($cm^3\ s^{-1}$)			
	15	30	60	90	15	30	60	90
Minimum	0.210	0.235	0.205	0.240	0.00008	0.00005	0.00009	0.00003
Maximum	0.308	0.333	0.306	0.342	0.00241	0.00208	0.00324	0.00856
Mean	0.245	0.272	0.230	0.280	0.00125	0.00114	0.00140	0.00167
Std. Error	0.018	0.020	0.024	0.024	0.00077	0.00078	0.00091	0.00220
Coefficient of variance	0.074	0.072	0.106	0.088	0.615	0.685	0.648	1.320

Temporal Variation of Soil Water Storage

Based on the monitoring of the SW characteristics, the soil volumetric water content (SVW) and soil water storage (SWS) in the top 30 and

90 cm were calculated, respectively (Fig. 3). The SVW and SWS of the 0–90 cm and 0–30 cm depth showed similar temporal trends. In the vertical dimension, the SVW of the same two depths (0–30 cm and 0–90 cm) had similar trends in most periods, except in the first month which revealed a difference due to thawing. The thawing process from the surface to the deep layer and the tillage in preparation for sowing in the top soil decreased the SVW at 0–30 cm depth. There was almost no difference in the SVW between the two depths from mid-May to August. Before the crop harvest period in September, the deeper soil had a higher SVW content than the surface layer due to less root activity and strong precipitation. The SWS of the 0–30 cm depth had a seasonal variation and had a more direct relationship with precipitation events. Before the rainfall on 1 July, the SWS remained at a stable level from mid-May to the end of June. The SWS had a periodic decrease when no rainfall occurred in late July. The SWS reached its summit with the large rainfall in the later of August.

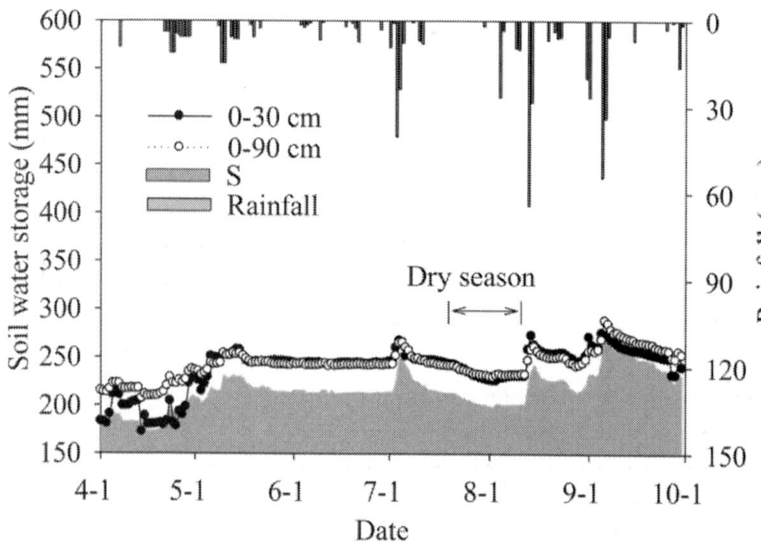

Figure 3: Temporal patterns of daily rainfall, soil water storage (SWS) and volumetric water content at 30 and 90 cm depths.

The SWS was calculated with the vertical variation in the SVW. The statistical characteristics of SWS and their proportions at each depth are listed in Table 3. The soil water storage averages at 0–20,

20–40 and 40–60 cm were 490.2, 593.8, and 358 m³ ha⁻¹. The vertical differences indicated the middle level has higher soil water content as a consequence of soil texture and crop root zone influences. Because precipitation was the only water source for dryland, the SWS at the three layers in the dry period was smaller than the average for the entire growing season, which also revealed the impact of maize ET on SWS. The SWS in the vertical dimension had the same trend in the entire period and the middle layer had the largest value (548.9 m³ ha⁻¹). The differences in SWS between the two periods become greater when going from the surface to the deeper layer, an observation derived from the rainfall contribution to the SWS in deeper layer.

Table 3: Dryland soil water storage and its proportion at three depths

	Duration		
Depth/ cm	Item	Entire growth season (30 May.-30 Sep.)	Dry season (20 Jul.-12 Aug.)
0-20	SWS/m³ ha⁻¹	490.2±34.6	441.1 ±19.5
	Proportion/%	23.4± 1.04	23.3±0.4
0-40	SWS/m³ ha⁻¹	1084±62.9	990.0±55.9
	Proportion/%	51.8±2.6	52.0±0.76
0-60	SWS/m³ ha⁻¹	1442±111	1303.0±70.6
	Proportion/%	68.8±1.6	68.5±0.55

Based on daily monitoring data, the accumulative patterns of evapotranspiration (ET), rainfall, and water loss in the growing season were compared (Fig. 4). The ET was at a low level in the first month of the growing period and then assumed a stable increasing rate beyond the 50th day after sowing. The accumulated SW loss was consistent with the fluctuation in the rainfall. The ET and rainfall stayed with the SW loss over the first 22 days. The accumulated rainfall amount increased by 79.8 mm; a large increase occurred on the 35th day when the ET and water loss increased slightly. The rainfall increased intensively after 80 days, which also caused an increase in the SW loss. The accumulated amounts of rainfall increased from 269.3 mm after 80 days to 440.9 mm after 120 days, and the SW loss increased from 47.9 mm after 75 days to 126.3 mm after 120 days. The temporal pattern demonstrated that the SW loss and ET of dryland were strongly correlated with the rainfall. The ET was also a maize growth feature and its similar pattern with rainfall also proved it was the dominant factor for crop growth.

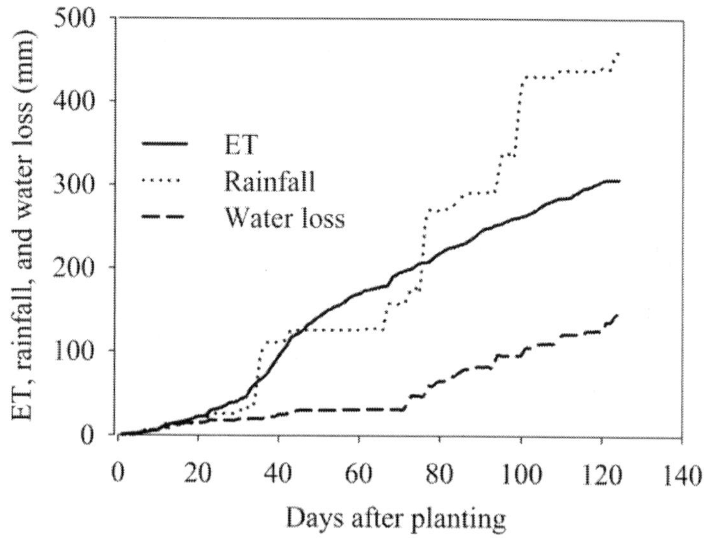

Figure 4: Cumulative evapo-transpiration (ET), rainfall, and water loss throughout maize growth period.

Variation of Crop Biomass in the Growing Period

The dry weight of the different maize organs (root, leaf, stem, tassel, and grain) in growing stages were analyzed (Fig. 5). Also shown in the Fig. 5 are the percentages of the total dry weight comprised by each organ in each of four growing seasons. The dry organ weight varied greatly from the seedling stage to maturity, a development period which is relatively short in this freeze-thawing area. The temporal pattern of dry weight in 2011 revealed the changes that occurred over four growing sections. The dry weight of grains increased significantly in the grain filling and maturity sections. The dry weight percentage did not change as intensively as the weight, and it was also found that the grains comprised nearly half of the maize weight in the maturity period.

With the N concentration in dried organs, the N absorption amount was quantified along with the organ dry weight. Due to logistical limitations, the crop organ sample in 2010 was only collected in the maturity period. The grain dry weights were larger in 2010 than in 2011. Comparing the dry weights of the organs in the maturity stage of the two years, the dry weights of the roots and tassel were significantly but their percentages differed little.

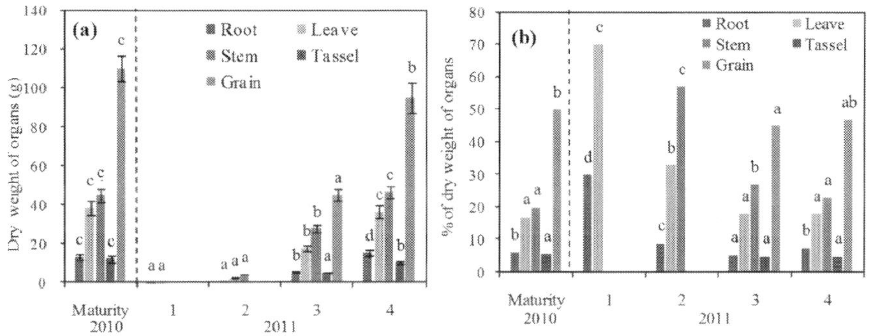

Figure 5: Dry weights (a) and percentage of maize organ weights (b) at different growth stages (Same letters in same organ between different growth stages indicate no significant difference.

1-Seedling, 2-Jointing, 3-Grain filling, 4-Maturity).

Nitrogen Uptake by Crop and Accumulation

The temporal N dynamics of crop uptake was then analyzed with the samples obtained from the maize organs (Fig. 6). As the maize grew, the N uptake in leaf first increased and then decreased with the maximum occurring in the jointing period. From grain filling to maturity, N uptake by leaf diminished insignificantly with a relative flat trend compared with that from the jointing to grain filling stages. As maize grew N uptake in stem decreased significantly decrease from 1 to 2 with a probable transfer of much N to tassel. The difference of N uptake by stem in the grain filling and maturity stages was insignificant, but the N uptake by grain increased significantly ($P<0.001$). The N uptake by leaf decreased insignificantly, which indicated that a large amount of N was transferred to tassel in the early time period and then transferred from tassel to grain (N uptakes by tassel in grain filling and maturity were 8.12 g kg^{-1} and 3.57 g kg^{-1}, respectively). There were insignificant differences in N uptakes of stem and leaf during the maturity stage in 2010 and 2011. The N uptake by grain in 2010 was 9.1% more than that in 2011.

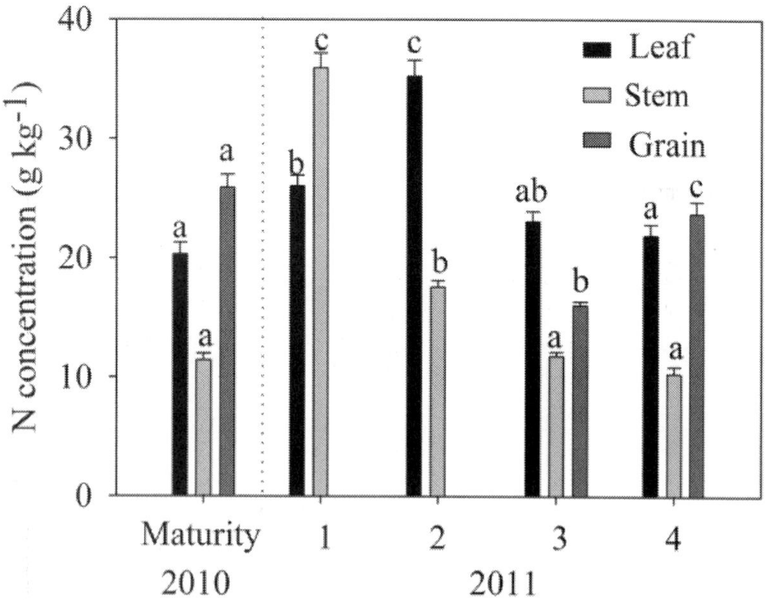

Figure 6: Seasonal dynamics of N concentration in maize organs (Same letters in same crop organs indicates no significant change between different growth stages.

1-Seedling, 2-Jointing, 3-Grain filling, 4-Maturity).

The accumulated N in different organs gradually increased in the early stages and reached a summit in the maturity period. From the jointing to grain filling period, the N in stem increased about 4.23 times and the amount in leaf increased less than five times. The N amounts in crop organs were not only related to the N uptake efficiency, but they also depended on the organ dry weight. According to the analysis in the maturity period, grain accumulated most of N, which occurred in a short time section. The total N amount of maize organs in 2010 was 309.9 kg ha^{-1}, which was larger than in 2011; the difference in grain was the dominant factor. The differences of N accumulation in leaf and stem during the maturity stages in the years 2010 and 2011 were insignificant.

Seasonal dynamics of N accumulative rates in organs during the growth period were analyzed (Fig. 7). From June to August, the N accumulative rates of stem and leaf increased rapidly and then declined. When the stem and leaf accumulated less N, the absorption efficiency in grain increased to 383 mg m^{-2}d^{-1}. The leaf was the most effective organ in uptaking N and the accumulation rate was about 500 mg m^{-2}d^{-1}.

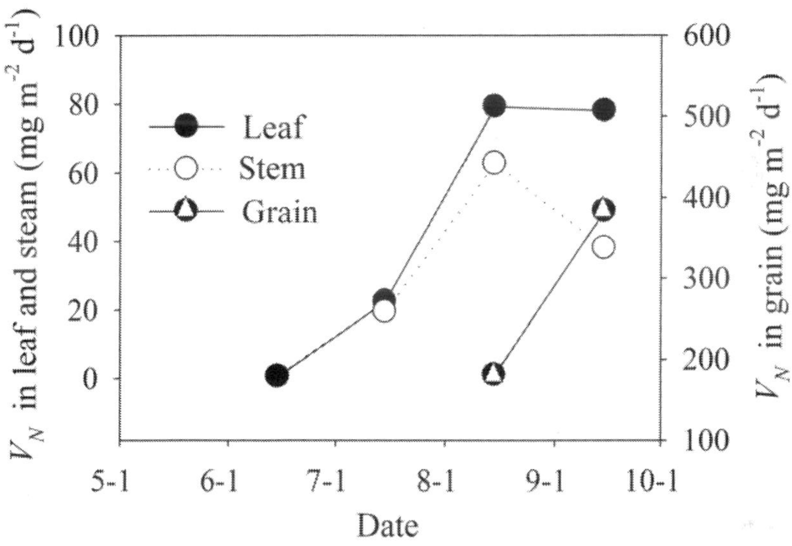

Figure 7: Seasonal dynamics of N accumulative rates (VN) in maize organs.

Variations of Soil N Stocks in the Growing Season

The N concentrations at four depths before sowing and after harvest were analyzed, which demonstrated the temporal-vertical soil N dynamics. The N (nitrate and ammonia) concentrations at different depths were first analyzed after the growing season (Table 5). Before the sowing, the ammonia-N stock at four depths ranged from 18.8 to 21.3 kg N ha^{-1}with no occurrence of a significant difference. After the harvest, the deeper layer experienced more loss than the surface layer. The middle two layers had close values and the 12.5 kg N ha^{-1} in top layer was the largest value. The nitrate-N concentration in soil was much larger than the ammonia-N. At the beginning of the growing season, the nitrate-N concentration was more than 121 kg ha^{-1} at all depths and had slight differences at 60–90 cm depth. The nitrate-N concentration decreased to 90 kg ha^{-1} after the harvest. The upper two layers had similar patterns and the concentrations ranged between 88–90 kg ha^{-1}. There was no significant difference in the concentration in the deeper two layers and their concentrations ranged from 63.2 to 68.0 kg ha^{-1}. The nitrate-N was more variable than ammonia-N and had a larger stock gap.

Table 4: Temporal pattern of N amount (kg ha^{-1}) of maize organs during growing season*

Year	Growth stage	Leaf	Stem	Grain
2010	Mature	58.6f	38.3ef	213de
2011	Seedling	0.14a		
	Jointing	6.85c	5.81a	
	Grain filling	30.6d	24.6d	54.1b
	Maturity	60.0f	36.0e	169c

Table 5: Mineral N stocks (kg ha^{-1}) and variations at each depth of growing season

Depth (cm)	Before sowing		After harvest		Stock gap	
	NO_3^--N	NH_4^+-N	NO_3^--N	NH_4^+-N	NO_3^--N	NH_4^+-N
0-15	147d	21.3b	90.1d	12.5c	-48.9c	-8.64bc
15-30	144d	21.2b	88.1d	8.12b	-53.8cd	-11.2c
30-60	130cd	19.3b	68.0c	9.02b	-65.3e	-9.79c
60-90	121c	18.8b	63.2c	5.74a	-56.8d	-12.69c

Nutrient Loss and Supply in Soil-Crop System

Based on field water monitoring and SW content in the 0–90 cm profile, the soil water loss (SL) and the leached N were calculated (Table 6). The rainfall and evapo-transpiration (ET) were calculated in the four stages of the maize growing season. The SL had a closer correlation with rainfall than with ET. The rainfall was nearly 100 mm in the jointing stage, but the SL was 11.5 mm due to the intensive ET. Moving from the grain filling stage to the maturity stage, the ET increased and more SL occurred under the higher rainfall event. The leached N was also calculated with the N concentration in soil water. The nitrate-N was the dominant N lost from the 90 cm soil profile, especially in the grain filling and maturity stages. The leached ammonia-N had similar temporal patterns and the largest amount was 0.18 kg ha^{-1}.

Table 6: Amounts of soil water loss (SL) and N leached below maize root zone (90 cm)

Growth stage	Period	R	ET	SL	leached N (kg ha^{-1})	
		(mm)			NO_3^--N	NH_4^+-N
Seedling	30 May to 29 Jun.	29.8	44.6	19.6	0.03	0.01
Jointing	30 Jun. to 25 Jul.	96.5	119.0	11.5	3.36	0.12
Grain filling	26 Jul. to 23 Aug.	165	67.3	48.6	5.53	0.18
Maturity	24 Aug. to 30 Sep.	168	76.6	67.5	6.26	0.10

Based on the N monitoring of rainfall and the survey of N fertilization, the N in rainfall and fertilizer was calculated. With the SWS and N concentration in SW in 0–90 cm profile, the *NUE*in the maize-soil system was calculated (Table 7). The N input from rainfall and fertilizer in the entire maize growing season was 16.5 and 120 kg ha^{-1}, respectively. The soil N supply in the entire growth period was about 641 kg ha^{-1}. With the nitrate and ammonia concentrations in SW and water volume, the N loss was estimated at 15.6 kg ha^{-1}. Based on the N balance, it was calculated that the *NUE* was 43%. On the other hand, the excessive N from dryland was an important potential source for N discharge to the environment.

Table 7: The N balance and usage efficiency (*NUE*) in crop-soil system during the maize growth period

Rainfall (kg N ha^{-1})	Fertilizer (kg N ha^{-1})	Initial soil N (kg N ha^{-1})		Final soil N (kg N ha^{-1})		N loss (kg N ha^{-1})		Soil supply N (kg N ha^{-1})	NUE (kg kg^{-1})
		NO_3^--N	NH_4^+-N	NO_3^--N	NH_4^+-N	NO_3^--N	NH_4^+-N		
16.5	120	542.9 ± 29.76	80.6 ± 4.29	309.4 ± 18.4	35.4 ± 4.37	15.2	0.41	641 ± 14.4	0.43 ± 0.03

DISCUSSION

The Dryland Soil Hydrological Process in a Freeze-Thawing Area

Rainfall is the only water source in dryland and it has a direct impact on SW content and storage [33]. The field monitoring done in four stages as part of our study indicated that the temporal patterns of SW content and acquisition rate closely followed the rainfall amount (Fig. 2). The soil roughness in a cultivated area is increased by tillage, which enhances the vertical transport of SW [34]. The SL was 146.7 mm (32% of rainfall in the growing season). However, the vertical distribution of SW content and soil water storage demonstrate that the water has a short vertical movement range. The SW can be sampled only after the rainfall events, but the acquisition rates showed slight vertical differences. These two findings both demonstrated that the vertical movement of water is slow even with large volumes and the speed can affect the range.

Compared with the long growing periods in other dryland agricultural areas, the local maize growing season is short and has a higher efficiency in the use of water and N [35]. A proper SW condition in the root zone is essential for crop growth in the maize seeding and grain filling stages. The SWS can remain at a stable level (about 1000 m^3 ha^{-1} of 40 cm profile) even in the dry season without precipitation (Fig. 3). All of these factors indicate that the *albic* soil texture can provide a positive water condition for crop growth [24]. The SW also controls surficial N movements in the dryland and provides a strong signal for crop absorption [36]. Crop transpiration will increase when the maize is near its maturity period, and the precipitation normally increases and meets this need during the same period (Fig. 4). The combined conditions of special soil texture and precipitation patterns directly determine the dryland soil hydrological pattern, which is the critical factor for maize growth in this freeze-thawing area.

Crop Growth and N Accumulation Differences in Plant Organs

The quantification and assessment of the organ biomass in crops can help in understanding the impacts of SW and N on crop growth. This information is also necessary for assessing N leaching [37]. The crop growth is measured by the organ dry weight (Fig. 5), which indicates the crop growth rate is higher than in other drylands under warming temperatures. The yearly precipitation in 2011 was more than in 2000 (Fig. 2), but did not lead to the higher grain dry weight in 2011 (Fig. 5). The difference proves that a higher maize yield is based on a combination of climate, soil and tillage. The grain filling period in this freeze-thawing area occurs in August, at a time when temperature and precipitation are at their highest levels. The active ET facilitates the transport of water and N from soil to crop, which finally causes a higher grain filling performance [38].

The absorbed N by maize organs in the entire growth period had different allocation rates (Fig. 7), which is determined by the photosynthesis of the organs and the eco-hydrological process. The crop accumulative rate in its canopy, stem and grain varied significantly, which means the N absorption capability for tissue production of dedicated organs shifts due to the N flux from soil to above ground biomass [39]. The soil N concentration in the vertical direction proved that the subsurface has a significant variance due to the root uptake. The accumulation rates of different organs over time also follow the typical sigmoid curve [40]. The temporal patterns of N in maize organs also indicate the N absorption is not only related to the soil N concentration, but that there is a complicated systemic feedback signal within the soil-crop system [41]. The SW availability, ET and N concentration are the key signaling factors. As the soil N is transported in crops with water, there are plenty of field observations and models to identify the growth stress reaction to the SW and N conditions [42],[43]. This information provides detailed guidelines for N and water adjustment in the soil-crop system.

Implication for N Leach Control and Soil N Optimization

The N uptake efficiency of a crop is the main component for *NUE* and the temporal patterns inTable 4 indicated that the highest amount of grain uptake of N occurs in the graining filling and maturity periods. The maize uptake of N will increase until the optimum levels of water and N in soils are reached [44]. Overlaying this information with the leached N loading, it was found that the intensive N uptake and loss occurred at same time. The strong ET and rainfall in that time period were the reason for the N transport in the crop-soil system [45]. The soil N gap analysis revealed that the observed N gap was similar in the vertical direction, which also demonstrates that the soil N can move among the soil depths with hydrological processes and maintain the N concentration balances [3]. The temporal N analysis in crop organs, soil and leached water explain the maize growth in this freeze-thawing area, which also provides a guideline for N fertilization while, also considering SW conditions.

The N pollution to groundwater due to excessive N fertilizer application is an important environmental concern [46]. The yearly *NUE* is about 43% in this area, which is a relatively high value in comparison with similar dryland agriculture areas [47]. The N leached from farmland mainly depends on the difference between the N obtained by the soil-water-crop system and the N uptake by the crop

[48]. As Table 6 shows, the leached N is mainly related to the loss of SW, which is the gap between rainfall and ET. The rainfall in August 2011 was above the long term monthly value, which caused the highest SW and N loss during the growing season. Under the normal climatic conditions, the *NUE* in this area can reach 50%. The nitrate-N is the dominant form of leached N to the groundwater, which is a direct consequence of fertilization and soil nitrification [49]. Models have been developed to optimize the fertilizer application to farmland based on a consideration of the soil hydrological process, which can help to maximize the crop yield with minimal discharge to the environment [50]. The model with local characteristic parameters, should be applied in the tillage practice to obtain an improved *NUE*.

CONCLUSION

The results of this study have shown that combined conditions of soil texture, soil hydrology and precipitation were basic factors for maize growth in the freeze-thawing area in Northeast China. The rainfall significantly affected the pattern of SW content and the allocation of N in the maize organs. Precipitation was the only water source for the dryland, but the variation in SW acquisition rates and SW content remained at a stable level due to the special *albic soil* texture. The dry weight and N accumulation of maize organs in the entire growing period proved that the soil hydrological condition was also enhanced.

The N loss was dependent on both the hydrological process and the N in the water, which was then decided by the rainfall water depth and the applied N. The NO_3-N was the dominant leachate, which was closely correlated with the amount of applied N fertilizer. The *NUE* was about 43%, which was also related to the initial soil N and maize uptake. In order to improve the*NUE* and decrease the amounts of leached nitrogen, it would be better to improve the fertilizer application amounts while giving consideration to rainfall patterns and N accumulation rates.

REFERENCES

1. Buckley C, Carney P (2013) The potential to reduce the risk of diffuse pollution from agriculture while improving economic performance at farm level. Environ Sci Policy 25(1): 118–126. doi: 10.1016/j.envsci.2012.10.002
2. Quemada M, Baranski M, Nobel-de Lange MNJ, Vallejo A, Cooper JM (2013) Meta-analysis of strategies to control nitrate leaching in irrigated agricultural systems and their effects on crop yield. Agric Ecosyst Environ 174(15): 1–10. doi: 10.1016/j.agee.2013.04.018
3. Ferrant S, Oehler F, Durand P, Ruiz L, Salmon-Monviola J, et al. (2011) Understanding nitrogen transfer dynamics in a small agricultural catchment: Comparison of a distributed (TNT2) and a semi distributed (SWAT) modeling approaches. J Hydrol 406(1–2): 1–15. doi: 10.1016/j.jhydrol.2011.05.026
4. Panda RK, Behera S (2003) Non-point source pollution of water resources: Problems and perspectives. J Food Agric Environ 1(3–4): 308–311.
5. Wang XB, Dai KA, Wang Y, Zhang XM, Zhao QS, et al. (2010) Nutrient management adaptation for dryland maize yields and water use efficiency to long-term rainfall variability in China. Agric Water Manage 97(9): 1344–1350. doi: 10.1016/j.agwat.2010.03.016
6. Yang YG, Xiao HL, Wei YP, Zhao LJ, Zou SB, et al. (2011) Hydrologic processes in the different landscape zones of Mafengou River basin in the alpine cold region during the melting period. J Hydrol 409(1–2): 149–156. doi: 10.1016/j.jhydrol.2011.08.013
7. Grzebisz W (2013) Crop response to magnesium fertilization as affected by nitrogen supply. Plant Soil 368(1–2): 23–39. doi: 10.1007/s11104-012-1574-z
8. Matejek B, Huber C, Dannenmann M, Kohlpaintner M, Gasche R, et al. (2010) Microbial N turnover processes in three forest soil layers following clear cutting of an N saturated mature spruce stand. Plant Soil 337(1–2): 93–110. doi: 10.1007/s11104-010-0503-2

9. Moreau P, Ruiz L, Vertes F, Baratte C, Delaby L, et al. (2013) CASIMOD'N: An agro-hydrological distributed model of catchment-scale nitrogen dynamics integrating farming system decisions. Agric Syst 118: 41–51. doi: 10.1016/j.agsy.2013.02.007
10. Kersebaum KC, Steidl J, Bauer O, Piorr HP (2003) Modelling scenarios to assess the effects of different agricultural management and land use options to reduce diffuse nitrogen pollution into the river Elbe. Phys Chem Earth 28: 537–545. doi: 10.1016/s1474-7065(03)00090-1
11. Hirel B, Le Gouis J, Ney B, Gallais A (2007) The challenge of improving nitrogen use efficiency in crop plants: towards a more central role for genetic variability and quantitative genetics within integrated approaches. J Exp Bot 58: 2369–2387. doi: 10.1093/jxb/erm097
12. Pask AJD, Sylvester-Bradley R, Jamieson PD, Foulkes MJ (2012) Quantifying how winter wheat crops accumulate and use nitrogen reserves during growth. Field Crop Res 126(14): 104–118. doi: 10.1016/j.fcr.2011.09.021
13. Bonato O, Schulthess F, Baumgärtner J (1999) Simulation model for maize crop growth based on acquisition and allocation processes for carbohydrate and nitrogen. Ecol Modell 124(1): 11–28. doi: 10.1016/s0304-3800(99)00146-5
14. Binder DL, Dobermann A, Sander DH, Cassman KG (2002) Biosolids as nitrogen source for irrigated maize and rainfed sorghum. Soil Sci Soc Am J 66(2): 531–543. doi: 10.2136/sssaj2002.0531
15. Favaretto N, Norton LD, Johnston CT, Bigham J, Sperrin M (2012) Nitrogen and Phosphorus Leaching as Affected by Gypsum Amendment and Exchangeable Calcium and Magnesium. Soil Sci Soc Am J 76(2): 575–585. doi: 10.2136/sssaj2011.0223
16. Lam QD, Schmalz B, Fohrer N (2012) Assessing the spatial and temporal variations of water quality in lowland areas, Northern Germany. J Hydrol 438–439(17): 137–147. doi: 10.1016/j.jhydrol.2012.03.011
17. Friend AD, Stevens AK, Knox RG, Cannell MGR (1997) A process-based, terrestrial biosphere model of ecosystem dynamics

(Hybrid v3.0). Ecol Modell 95(2–3): 249–287. doi: 10.1016/s0304-3800(96)00034-8

18. Dorioz JM, Ferhi A (1994) Non-point pollution and management of agricultural areas: Phosphorus and nitrogen transfer in an agricultural watershed. Water Res 28(2): 395–410.
19. Keeney DR, Follett RF (1991) Managing Nitrogen for Groundwater Quality and Farm Profitability, SSSA, Madison, WI. 1–7.
20. Kim DG, Vargas R, Bond-Lamberty B, Turetsky MR (2012) Effects of soil rewetting and thawing on soil gas fluxes: a review of current literature and suggestions for future research. Biogeoscience 9(7): 2459–2483. doi: 10.5194/bg-9-2459-2012
21. Lewis DR, McGechan MB, McTaggart IP (2003) Simulating field-scale nitrogen management scenarios involving fertiliser and slurry applications. Agr Syst 76(1): 159–180. doi: 10.1016/s0308-521x(02)00032-x
22. Dawson JC, Huggins DR, Jones SS (2008) Characterizing nitrogen use efficiency in natural and agricultural ecosystems to improve the performance of cereal crops in low-input and organic agricultural systems. Field Crop Res 107: 89–101. doi: 10.1016/j.fcr.2008.01.001
23. Schroder JL, Zhang HL, Girma K, Raun WR, Penn CJ, et al. (2011) Soil Acidification from Long-Term Use of Nitrogen Fertilizers on Winter Wheat. Soil Sci Soc Am J 75(3): 957–964. doi: 10.2136/sssaj2010.0187
24. Caliskan S, Ozkaya I, Caliskan ME, Arslan M (2008) The effects of nitrogen and iron fertilization on growth, yield and fertilizer use efficiency of soybean in a Mediterranean-type soil. Field Crop Res 108(2): 126–132. doi: 10.1016/j.fcr.2008.04.005
25. Razzaghi F, Plauborg F, Jacobsen SE, Jensen CR, Andersen MN (2012) Effect of nitrogen and water availability of three soil types on yield, radiation use efficiency and evapotranspiration in field-grown quinoa. Agric Water Manage 109: 20–29. doi: 10.1016/j.agwat.2012.02.002
26. Hao FH, Chen SY, Ouyang W (2013) Temporal rainfall patterns with water partitioning impacts on maize yield in a freeze-thaw zone. J Hydrol 486(12): 412–419. doi: 10.1016/j.jhydrol.2013.02.008

27. Ouyang W, Huang HB, Hao FH, Shan YS, Guo BB (2012) Evaluating spatial interaction of soil property with non point source pollution at watershed scale: the phosphorus indicator in Northeast China. Sci Total Environ 432: 412–421. doi: 10.1016/j.scitotenv.2012.06.017
28. Jackson JE (1991) A user's guide to principal components. Wiley-Interscience, New York.
29. Moroizumi T, Hamada H, Sukchan S, Ikemoto M (2009) Soil water content and water balance in rainfed fields in Northeast Thailand. Agric Water Manage 96(1): 160–166. doi: 10.1016/j.agwat.2008.07.007
30. Huggins DR, Pan L (1993) Nitrogen efficiency component analysis: an evaluation of cropping system differences in productivity. Agronomy J 85: 898–905. doi: 10.2134/agronj1993.00021962008500040022x
31. Meisinger JJ, Randall GW (1991) Estimating nitrogen budgets for soil-crop systems. In: Follet, R.F., et al. (Eds.), Managing Nitrogen for Groundwater Quality and Farm Profitability. SSSA, Madison,WI, USA, 85–124.
32. Vázquez N, Pardo A, Suso ML, Quemada M (2006) Drainage and nitrate leaching under processing tomato growth with drip irrigation and plastic mulching. Agric Ecosyst Environ 112(4): 313–323. doi: 10.1016/j.agee.2005.07.009
33. Tanveer SK, Wen XX, Lu XL, Zhang JL, Liao YC (2013) Tillage, Mulch and N Fertilizer Affect Emissions of CO2 under the Rain Fed Condition. PLOS ONE 8(9): e72140. doi: 10.1371/journal.pone.0072140
34. Dörner J, Horn R (2009) Direction-dependent behaviour of hydraulic and mechanical properties in structured soils under conventional and conservation tillage. Soil Till Res 102(2): 225–232. doi: 10.1016/j.still.2008.07.004
35. Murungu FS, Chiduza C, Muchaonyerwa P, Mnkeni PNS (2011) Mulch effects on soil moisture and nitrogen, weed growth and irrigated maize productivity in a warm-temperate climate of South Africa. Soil Till Res 112(1): 58–65. doi: 10.1016/j.still.2010.11.005

36. Gheysari M, Mirlatifi SM, Homaee M, Asadi ME, Hoogenboom G (2009) Nitrate leaching in a silage maize field under different irrigation and nitrogen fertilizer rates. Agric Water Manage 96(6): 946–954. doi: 10.1016/j.agwat.2009.01.005
37. Wegehenkel M, Mirschel W (2006) Crop growth, soil water and nitrogen balance simulation on three experimental field plots using the Opus model—A case study. Ecol Modell 190(1-2): 116–132. doi: 10.1016/j.ecolmodel.2005.02.020
38. Katerji N, Campi P, Mastrorilli M (2013) Productivity, evapotranspiration, and water use efficiency of corn and tomato crops simulated by AquaCrop under contrasting water stress conditions in the Mediterranean region. Agric Water Manage 130: 14–26. doi: 10.1016/j.agwat.2013.08.005
39. Strullu L, Cadoux S, Preudhomme M, Jeuffroy MH, Beaudoin N (2011) Biomass production and nitrogen accumulation and remobilisation by Miscanthus × giganteus as influenced by nitrogen stocks in belowground organs. Field Crop Res 121(3): 381–391. doi: 10.1016/j.fcr.2011.01.005
40. Shoji S, Gandeza AT, Kimura K (1991) Simulation of crop response to polyolefin-coated urea: II. Nitrogen uptake by corn. Soil Sci Soc Am J 55(5): 1468–1473. doi: 10.2136/sssaj1991.03615995005500050045x
41. Nacry P, Bouguyon E, Gojon A (2013) Nitrogen acquisition by roots: physiological and developmental mechanisms ensuring plant adaptation to a fluctuating resource. Plant Soil 370(1–2): 1–29. doi: 10.1007/s11104-013-1645-9
42. Kovács GJ (2005) Modelling of adaptation processes of crops to water and nitrogen stress. Phys Chem Earth, Parts A/B/C 30(1–3): 209–216. doi: 10.1016/j.pce.2004.08.022
43. Overman AR, Scholtz RV (2013) Accumulation of Biomass and Mineral Elements with Calendar Time by Cotton: Application of the Expanded Growth Model. PLOS ONE 8(9): e72810. doi: 10.1371/journal.pone.0072810
44. Kerbiriou PJ, Stomph TJ, Van Der Putten PEL, Van Bueren ETL, Struik PC (2013) Shoot growth, root growth and resource capture under limiting water and N supply for two cultivars of lettuce (Lactuca sativa L.). Plant Soil 371(1–2): 281–297. doi: 10.1007/s11104-013-1672-6

45. Asseng S, Turner NC, Keating BA (2001) Analysis of water- and nitrogen-use efficiency of wheat in a Mediterranean climate. Plant Soil 233(1): 127–143. doi: 10.1023/a:1010381602223

46. Fang QX, Yu Q, Wang EL, Chen YH, Zhang GL, et al. (2006) Soil nitrate accumulation, leaching and crop nitrogen use as influenced by fertilization and irrigation in an intensive wheat-maize double cropping system in the North China Plain. Plant Soil 284(1–2): 335–350. doi: 10.1007/s11104-006-0055-7

47. Ouyang W, Wei XF, Hao FH (2013) Long-term soil nutrient dynamics comparison under smallholding land and farmland policy in northeast of China. Sci Total Environ 450–451(15): 129–139. doi: 10.1016/j.scitotenv.2013.02.016

48. Hall JA, Bobe G, Hunter JK, Vorachek WR, Stewart WC, et al. (2013) Effect of Feeding Selenium-Fertilized Alfalfa Hay on Performance of Weaned Beef Calves. PLOS ONE 8(3): e58188. doi: 10.1371/journal.pone.0058188

49. Lteif A, Whalen JK, Bradley RL, Camire C (2010) Nitrogen transformations revealed by isotope dilution in an organically fertilized hybrid poplar plantation. Plant Soil 333(1–2): 105–116. doi: 10.1007/s11104-010-0326-1

50. Sierra J, Brisson N, Ripoche D, Noel C (2003) Application of the STICS crop model to predict nitrogen availability and nitrate transport in a tropical acid soil cropped with maize. Plant Soil 256(2): 333–345.

Chapter 5

Two Challenges for U.S. Irrigation Due to Climate Change: Increasing Irrigated Area in Wet States and Increasing Irrigation Rates in Dry States

Robert I. McDonald[1*], Evan H. Girvetz[2]

[1]Central Science Program, The Nature Conservancy, Arlington, Virginia, United States of America,

[2]Global Climate Change Program, The Nature Conservancy, Seattle, Washington, United States of America

ABSTRACT

Agricultural irrigation practices will likely be affected by climate change. In this paper, we use a statistical model relating observed water use by U.S. producers to the moisture deficit, and then use this statistical model to project climate changes impact on both the fraction of agricultural land irrigated and the irrigation rate (m^3ha^{-1}). Data on water withdrawals for US states (1985–2005) show that both quantities are highly positively correlated with moisture deficit (precipitation – PET). If current trends hold, climate change would increase agricultural demand for irrigation in 2090 by 4.5–21.9 million ha (B1 scenario demand: 4.5–8.7 million ha, A2 scenario demand: 9.1–21.9 million ha). Much of this new irrigated area would occur in states that currently have a wet climate and a small fraction of their agricultural land currently irrigated, posing a challenge to policymakers in states with less experience with strict regulation of agriculture water use. Moreover, most of this expansion will occur in states where current agricultural production has relatively low market value per hectare, which may make installation of irrigation uneconomical without significant changes in crops or practices by producers. Without significant increases in irrigation efficiency, climate change would also increase the average irrigation rate from 7,963 to 8,400–10,415 m^3ha^{-1} (B1 rate: 8,400–9,145 m^3ha^{-1}, A2 rate: 9,380–10,415 m^3ha^{-1}). The irrigation rate will increase the most in states that already have dry climates and large irrigation rates, posing a challenge for water supply systems in these states. Accounting for both the increase in irrigated area and irrigation rate, total withdrawals might increase by 47.7–283.4 billion m^3 (B1 withdrawal: 47.7–106.0 billion m^3, A2 withdrawal: 117.4–283.4 billion m^3). Increases in irrigation water-use efficiency, particularly by reducing the prevalence of surface irrigation, could eliminate the increase in total irrigation withdrawals in many states.

INTRODUCTION

Anthropogenic emissions of greenhouse gases are very likely increasing average temperatures and are more likely than not altering the amount and timing of precipitation [1]. The effects of climate change on agriculture will likely be multifaceted [2]–[5]. Changes in temperature, precipitation [6] and CO_2 concentration effects will affect plant growth. Moreover, the indirect effects of changes in weed and pest abundance and disease outbreaks may also significantly affect agriculture [7], [8]. This paper focuses on irrigation, the largest human water use, which will likely be affected by climate change [9].

Climate change will likely affect the supply of water available from surface water or groundwater, decreasing it in many cases, and may also increase evapotranspiration from crops and hence demand for irrigation [10]–[16]. While several papers have quantified how changes in climate will affect irrigation water use by agricultural producers, results have been mixed depending on the time horizon of the analysis, the particular general circulation models (GCMs) consulted, and assumptions about how producers' irrigation practices respond to climate change [17]–[20].

The decision of whether or not to irrigate an agricultural field, as well as the amount of irrigation water applied, is a complex decision for producers, and depends on a number of factors [18],[21]. Water supply is constrained by water policy and law as well as the infrastructure available to bring water to farmers. The potential economic return from production depends on the crops grown, the productivity of the soil, and the other inputs farmers add such as fertilizer. Demand for irrigation also depends on the crop grown, as well as the prevailing climate. Climate change might increase moisture deficit, or decrease it, depending on the relative change in temperatures and precipitation. Accounting for all these factors is difficulty, which makes forecasting irrigation water use challenging.

This paper uses an empirical approach, showing for the United States that changes in climate and irrigation equipment over the past 25 years have resulted in predictable changes in both the fraction of agricultural land irrigated and the irrigation rate (m^3 ha^{-1}). We then use a simple statistical model to make state-level projections of how both quantities will be affected by climate change. We incorporate information on historical changes over time in irrigation equipment into our statistical model, and project the potential scope of future irrigation efficiency improvements to serve as an adaptation to climate change. We use this analysis to estimate the how irrigation rate and area irrigated will need to change in the future–based on the assumptions of this model–to keep up with the change in water demand of crops due to climate change.

MATERIALS AND METHODS

Data

Data on historical withdrawals for the period 1985–2005 were obtained from the United States Geological Survey (USGS), which conducts county-level surveys every five years [22]. For this study, we use total freshwater withdrawals (surface water plus groundwater) by the agriculture sector. For our analysis, we lump county-level data to states, since the county-level data vary widely in withdrawals when the county in which an irrigation system withdraws water is different than the county where the water is applied.

Withdrawal information was supplemented by data on area irrigated, as defined in the United States Department of Agriculture (USDA) Farm and Ranch Irrigation Survey [23], and total agricultural area and its market value, as defined in the USDA Census of Agriculture [24]. In this study, our definition of "agricultural area" follows the USDA definition of "total cropland", which includes harvested cropland, cropland used for pasture or grazing, and other miscellaneous cropland (e.g., areas where crops failed, fields were left intentionally left fallow, or with cover crops). Note that the subcategory of "cropland used for pasture and grazing" includes only land currently used for pasture or grazing that could be immediately used for crops without any additional improvement; it does not include the much larger land area of woodland and rangeland that is used for grazing or pasturing.

Data on historical temperature and precipitation were taken from the Parameter-elevation Regressions on Independent Slopes Model (PRISM) dataset (~4 km resolution) and used to calculate historic observed moisture deficit [25]. Future climate scenarios are based on the ensemble mean of a panel of 16 general circulation models (GCM) [26], [27] for each of the A2, A1B, and B1 scenarios of the Intergovernmental Panel on Climate Change [28]. The 16 GCMs used in this study are BCCR-BCM2.0, CGCM3.1(T47), CNRM-CM3, CSIRO-Mk3.0, GFDL-CM2.0, GFDL-CM2.1, GISS-ER, INM-CM3.0, IPSL-CM4, MIROC3.2(medres), ECHO-G, ECHAM5/MPI-OM, CCSM3, PCM, and UKMO-HadCM3. For more information on the structure of each GCM, see Meehl et al. [26]. We examined three time periods:

2020–2039 (hereafter "2030"), 2040–2059 (hereafter "2050"), 2060–2079 (hereafter "2070"), and 2080–2099 (hereafter "2090"). To capture uncertainty in the GCM predictions of climate under different emissions scenarios, we also include the 20% and 80% quintiles of the ensemble's prediction of moisture deficit.

By using an ensemble of 16 GCMs, and considering the variation among the ensemble predictions for GCMs, we hope to improve on previously published work by considering a broad set of GCMs. Other papers that have quantified how changes in climate will affect irrigation water use have only considered a few GCMs, and some of the variation in the results among and within the papers seems to be driven by the particular GCMs chosen [17]–[20]. For instance, Fisher et al. [17] looked at the climate change impacts of the a2r emissions scenario with the HadCM3 and CISRO GCMs, over the time period 1990–2080, projecting a 45% increase in irrigated area globally. Their predictions for North America varied between the two GCMs, because the CSIRO MK3 predicts a decrease in moisture deficit for much of the US, while the Hadley CM3 predicts an increase in moisture for much of the U.S. Similarly, Thomson et al. [19] looked at the climate change impacts of different average temperature increases with two GCMs, the HadCm3 and CGCM, but incorporated in their model both demand for irrigation and supply of available water. They predict that irrigated area will decline in the continental United States under both GCMs, despite strong growth in demand, because of a decline in the available supply of irrigation water.

Calculation of Moisture Deficit

We calculated the soil moisture deficit during the growing season by subtracting precipitation from potential evapotranspiration (PET), as measured with the modified Thornthwaite (Hamon) method [29], which is based on temperature and number of daylight hours, and widely used in global and regional hydrologic models [30]. This metric should theoretically be related to irrigation water required. Moreover, it integrates intra-annual variability in water demand into a cumulative number for the growing season.

GCM predictions of moisture deficit were first resampled to average county values, and then average state values were calculated, weighting by the agricultural area (for the analysis of fraction of area irrigated) or irrigated area (for the analysis of irrigated area).

There has been considerable debate about the merits of different methods of estimating PET, and Kingston et al. [31] showed that the relative magnitude of the effect of climate change on PET depends on which method is used. The Hamon method appears to be one of the more sensitive metrics, increasing with climate change greatly, whereas other methods like Blaney-Criddle and Priestley-Taylor increase less [31]. To ensure that our results were not influence by our choice of metrics, we also calculated PET using the Blaney-Criddle method, a simple method that is known to be relatively less sensitive to climate change than most metrics [31]. The Blaney-Criddle method is a function solely of the mean daily temperature and the mean daily percentage of daytime hours.

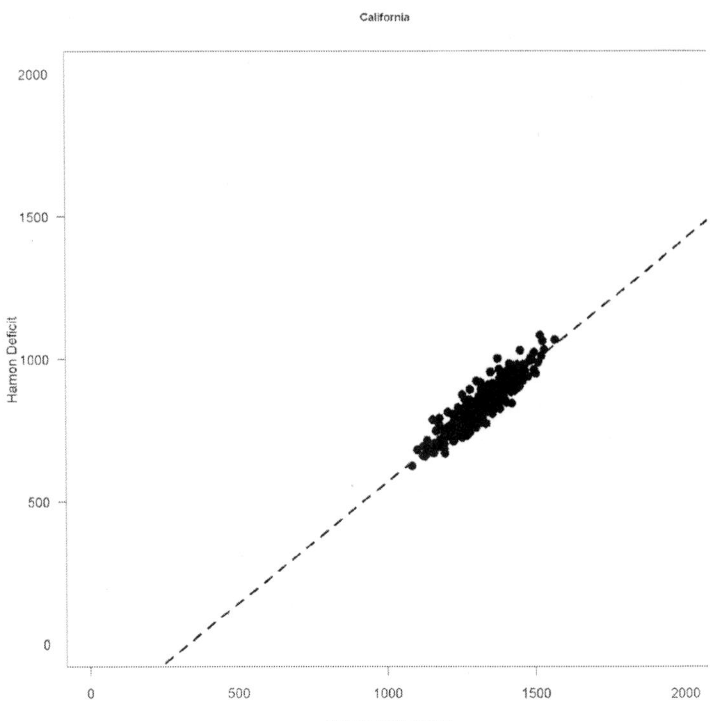

Figure 1: Scatterplot of estimated PET for California in 2005, using the Hamon and Blaney-Criddle metrics.

Preliminary results suggest no major change in our findings. To see why this is likely to be so, note that we build a statistical model that relates irrigation use by farmers to historical values of PET. Since Hamon and Blaney-Criddle give highly correlated estimates of PET (Figure 1), and changes in Hamon and changes in Blaney-Criddle are highly correlated (Figure 2), they both give similar results in the linear regression we use as our statistical model. While change in PET as estimated by Hamon is systematically higher than change in PET by Blaney-Criddle, the fitted slopes in our regression correct for this fact and given similar estimates of irrigation use. Because we are fitting either estimate of PET to empirical data on irrigation use, as long as the estimates are strongly linearly correlated with one another, the projections of the statistical model will be very similar.

Each dot represents one year in a particular combination of GCM and greenhouse gas emissions Scenario. Note that the strong linear correlation between the two means that when either of these two metrics are statistically related to irrigation use, the quantitative predictions of the effect of climate change are quite similar. Other states also show a linear correlation, with R^2 ranging from 0.75 to 0.93.

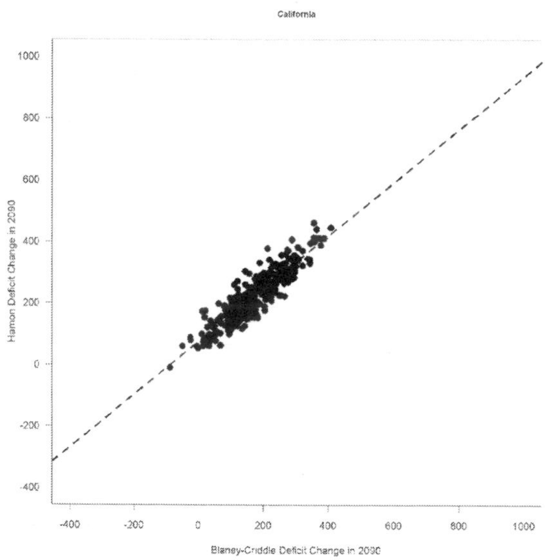

Figure 2: Scatterplot of change in PET for California between 2005 and 2090, using the Hamon and Blaney-Criddle metrics.

Each dot represents one year in a particular combination of GCM and greenhouse gas emissions scenario. Note that the strong linear correlation between the two means that when either of these two metrics are statistically related to irrigation use, the quantitative predictions of the effect of climate change are quite similar. Other states also show a linear correlation, with R^2 ranging from 0.74 to 0.93.

The spatial patterns of change in Hamon and Blaney-Criddle are quite similar (Figure 3). However, there are slight differences among states. For example, the Hamon metric predicts higher increases in PET in arid parts of Arizona than does Blaney-Criddle.

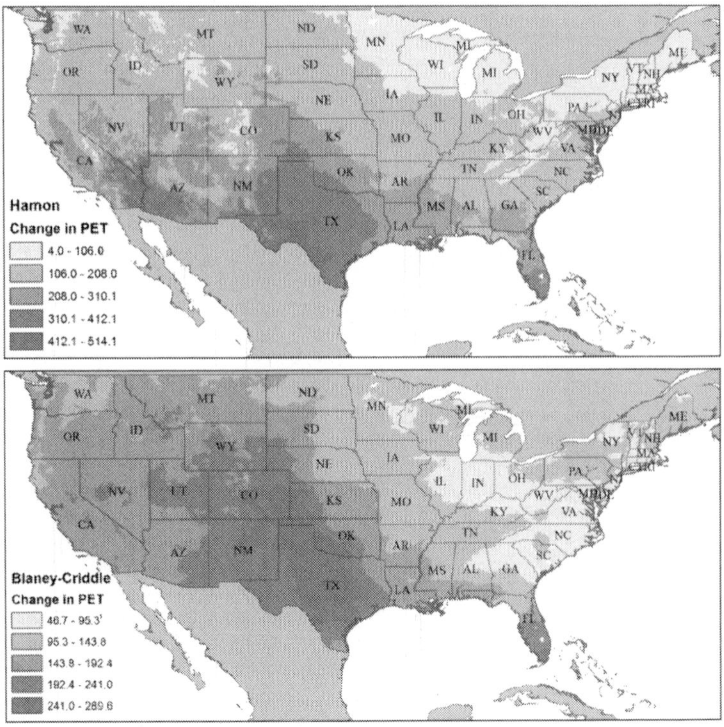

Figure 3: The effect of climate change on the Hamon estimate of PET (top) and the Blaney-Criddle estimate of PET (bottom).

Both panels are colored with 5 equal interval categories that linearly span the range of the pixel values, with areas of less increase in PET being yellow and areas of greater increase in PET being red.

Statistical Analysis

Total withdrawals were modeled as a function of total agricultural land (*A*) times the fraction of agricultural land that is irrigated (*F*) times the irrigation rate (*R*):

$$W_t = A_{tot,t} F_t R_t$$

Historical information is taken from the USDA survey. *F* and *R* were both modeled as functions of moisture deficit in 1985 and the change in moisture deficit ($\text{Deficit}_t - \text{Deficit}_{1985}$):

$$\pi_0 + \pi_1 \text{Deficit}_{1985} + \pi_2 \text{Deficit}_\Delta$$

where

$$\pi_0 = \mu_0 + \alpha_{t,0} + \beta_0 \text{Flood}_t$$

$$\pi_1 = \mu_1 + \alpha_{t,1} + \beta_1 \text{Flood}_t$$

The first two terms (π_0 and π_1) can be interpreted as, respectively, the slope and intercept of the relationship between moisture deficit and the response variable at a point in time. This relationship is the long-term average relationship with climate, and is the product of decades of economic decisions and irrigation policies. The third term (π_2) is the effect on the response variable of an inter-annual change in moisture deficit. It is the short term response of the agricultural sector to an interannual change in moisture deficit. All three terms are a function of a grand mean (μ), a time-specific shock (α_t), and the percent of surface irrigation (*Flood*). Surface irrigation (irrigation techniques where water is applied to the soil surface using gravity) was selected because it was believed to be causally negatively related to irrigation water-use efficiency. In this dataset proportion surface plus proportion sprinkler is approximately 1, since microirrigation is relatively uncommon. Note that the time-specific shock terms (α) will also incorporate other factors that varied over time but are not explicitly modeled here. For instance, if the steadily rising ambient CO_2 concentrations increased crop water-use efficiency over the observed 20 year time period, then one would expect the $\alpha_{t,1}$ to decline over time, all else being equal.

To account for autocorrelated errors by state, we used a repeated measured design for the error term, using Proc Mixed in SAS, version 9.2. We added terms one by one to the model, stopping when the added variables did not significantly improve the negative log-likelihood. The functional form of the regression of *F* was a logistic regression, although model fits using other functional forms that are bounded between 0 and 1 (e.g., probit) yielded similar results. For the regression of *R*, irrigation rates were log transformed to improve normality and to bound our results to the positive domain.

Future projections were calculated by assuming the fitted parameters in the regression equations, based on historical data, continue into the future. This implicitly assumes that some adaptation by producers takes place: regions that dry out, for example, adapt the agricultural practices of current comparable dry regions. Of course, hypothetically much larger water savings (or waste) are possible with more (or less) drastic change in agricultural practices, but we do not consider these more extreme possibilities in this analysis. Note that our analysis ignores the potential effect of increased CO_2 concentrations increasing crop water-use efficiency, although we highlight the importance of this issue in the Discussion section.

RESULTS

Analysis of Observed Historic Irrigation Patterns

During 1985–2005 the proportion of irrigated agricultural land was greater in areas with greater moisture deficit ($p<0.001$, Table 1; and Figure 4). In any year, states with higher moisture deficits have greater proportions of agricultural land under irrigation. This relationship represents the long-term outcome of individual farmers' decisions as a function of climate. Note that this long-term historical relationship shows adaptation to historical climate, in that dry areas are growing different crops and using different practices than wet areas. On the shorter year-to-year time scale, inter-annual changes in moisture deficit have resulted in changes to the proportion of agricultural area irrigated: higher moister deficit resulted in higher proportion of area irrigated (p = 0.03, Table 1).

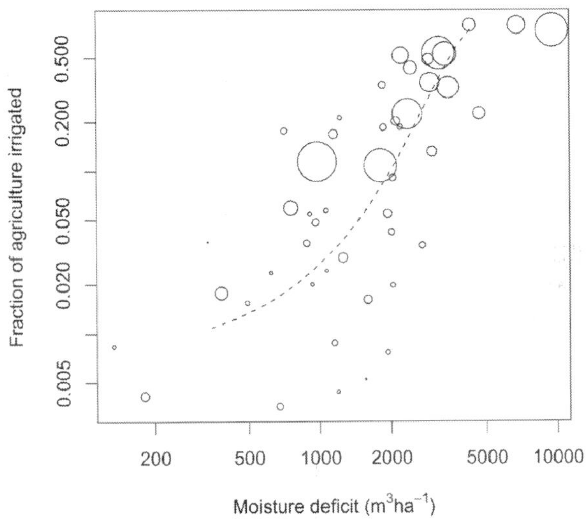

Figure 4: The fraction of agricultural land irrigated in U.S. states in 2005 as a function of moisture deficit.

Note the logarithmic scale on both axes. The size of circle is proportional to the irrigation rate (m^3ha^{-1}); states with high moisture deficit have higher irrigation rates. The fitted regression line is shown for the middle 90% of the data range.

Table 1: Regression coefficients that predict the proportion of agricultural area that is irrigated, logit transformed

Effect	Estimate	SE	P
Intercept (μ_0)	-4.4049	0.1798	<.0001
Deficit1985 (μ_1)	0.000807	0.000067	<.0001
$Deficit_{\Delta}$ (μ_2)	0.000371	0.000164	0.0252

Similarly, the irrigation rate (m^3 of water per hectare of irrigated land) is greater in sites with greater moisture deficit ($p<0.001$, Table 2 and Figure 4). At any point in time, states with higher moisture deficits have higher irrigation rates. Again, this relationship among different sites is the long-term outcome of a historical process of agricultural development that was influenced by the prevailing climate, and the gradient occurs despite some presumed adaptation of agricultural crops and practices to historical climate. Interestingly, the historical shift away from surface irrigation and toward sprinkler and drip irrigation has resulted in decreased irrigation rates over time ($p<0.001$, Table 2), all else being equal. In the short run on the year-to-year time scale, increases in moisture deficit are associated with increased irrigation rate, although this result is not statistically significant ($p = 0.11$, Table 2).

Table 2: Regression coefficients that predict the irrigation rate (m^3ha^{-1}), log transformed

Effect	Estimate	SE	P
Intercept (μ_0)	7.7495	0.07894	<.0001
Deficit1985 (μ_1)	0.000132	0.000037	0.0009
Deficit Δ (μ_2)	0.000118	0.000073	0.11
Fraction Flood1985 (β_1)	2.1475	0.4675	<.0001
Fraction Flood Δ (β_2)	1.3409	0.1988	<.0001

Projected Impact of Climate Change on Irrigation Area

Moisture deficit is projected to increase on average for US states under all three climate change scenarios (Figure 5A). After 2060, the increase in moisture deficit is largest in the A2 scenario, and smallest in the B1 scenario. Regardless of emissions scenario, climate change increases moisture deficit in 2090 throughout most of the contiguous US for 13 of the 16 GCMs considered [27], with the exception of the NCAR CCSM3, CSIRO MK3, and NCAR PCM models. Generally, the largest increase in moisture deficit is in the South followed by the Southwest, while the Northeast has small increases or decreases in moisture deficit. The most disagreement among climate models in the direction of moisture deficit change is in the Southwest, where some GCMs project a decrease and some an increase in moisture deficit.

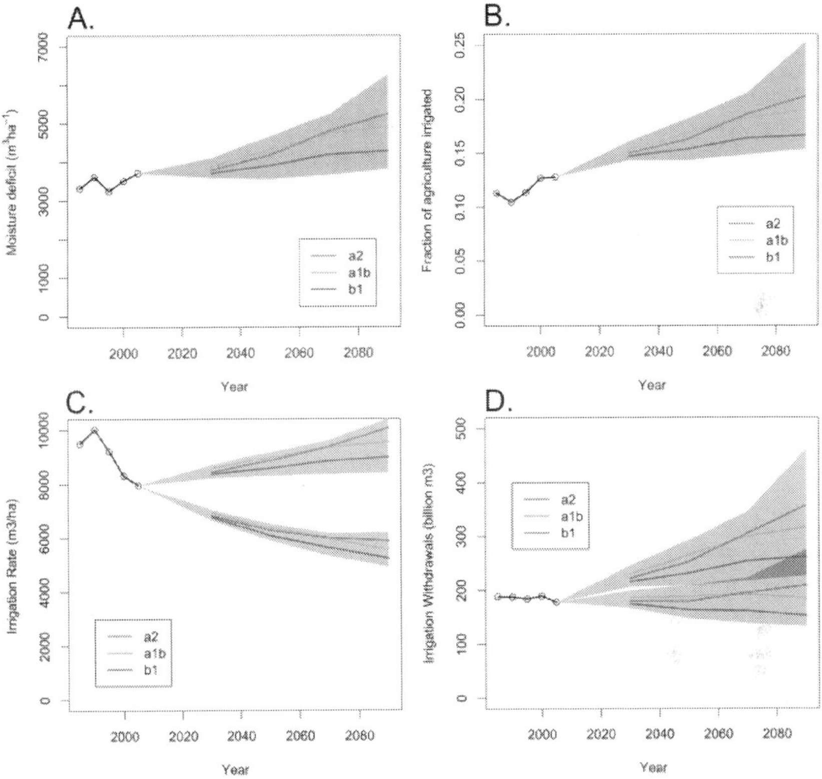

Figure 5: The effect of climate change on irrigation water use by United States agriculture.

A.) Historical and future trends in U.S. mean moisture deficit. Displayed is the irrigated area weighted average, which is most directly relevant to irrigation rate. For each climate change scenario, the projected moisture deficit at the climate of the ensemble median is shown. The grey area shows the range of fraction irrigated for climate of the 20th and 80th quintiles of the ensemble. B.) Historical and future trends in the fraction of agriculture irrigated. C.) Historical and future trends in the irrigation rate. The blue area shows the effect of climate change on irrigation rates if the current mix of irrigation equipment persists over time; the green area shows the effect of climate change if the observed (1985–2005) trend away from relatively inefficient surface irrigation continues over time. D.) Historical and future trends in total irrigation withdrawals. Note the confidence intervals of the blue (current mix of irrigation equipment) and green (decreased use of surface irrigation) areas overlap after 2030.

Given the projected increases in moisture deficit with climate change, we project that the proportion of agricultural land irrigated in the U.S. will increase (Figure 5B). Under all emissions scenarios, farmers will increasingly need to irrigate their fields as it gets drier. The total area irrigated nationally will increase by 4.5–21.9 million ha, with the uncertainty due both to differences among emissions scenarios and to differences among GCMs. The B1 emissions scenario has the smallest increase (4.5–8.7 million ha) and the A2 emissions scenario has the greatest (9.1–21.9 million ha). Note that our projection estimates new irrigated area based on observed trends, after a moderate degree of adaptation: our methodology implicitly assumes farmers in a site made drier by climate change adopt the crops and practices of current farmers in comparable climates. Depending on the state, laws governing access to water or a shortage of available water might prevent this much expansion of irrigated area from occurring [18]. Alternatively, failure of farmers to adapt to climate change as much as is assumed in our model would imply an even larger increase in irrigated area.

The states with the greatest demand for new irrigated area, in ha, are those in the South-Central U.S. (Figure 6A) and to a lesser extent the Northwest and California. There is projected to be less of an increase in the Northeast, where climate impacts on moisture deficit are projected to be smaller. The extent of expansion of irrigation is a function of the amount of greenhouse gas emissions, with the A2 scenario requiring the most expansion and the B1 scenario the least. The difference in projected new irrigation area demanded between the A2 scenario and the B1 scenario, at the ensemble median climate projections, is 6.4 million ha. While exact cumulative emissions depend on the GCM implementation of the scenarios, every 1 GtC increase in cumulative emissions increases the demand for new irrigated land in the US by roughly 7000 ha.

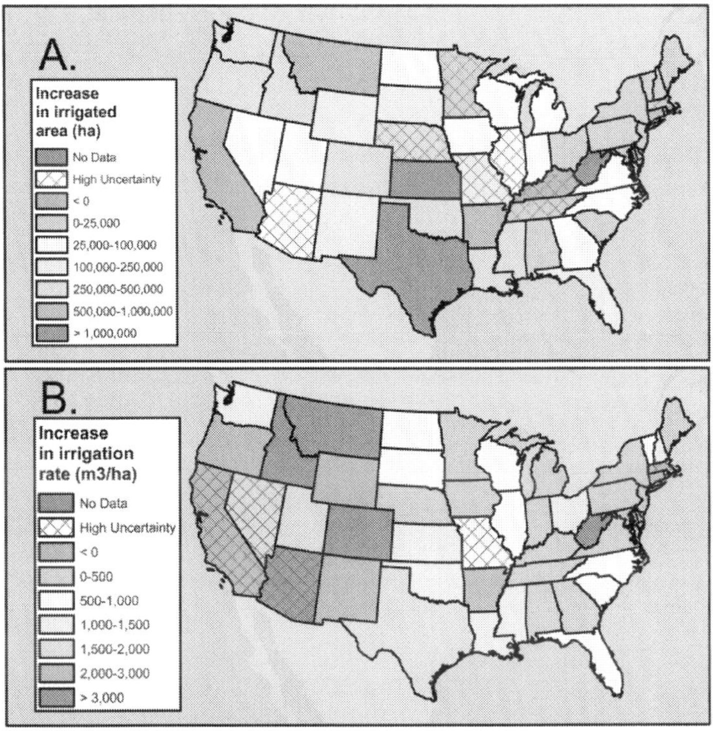

Figure 6: The effect of climate change on irrigation by states. A.) Projected increase in irrigated area by 2090 under the A1B scenario, ensemble median. Most states have an increase in irrigated area under all emission scenarios in more than 80% of the GCMs in the ensemble; those that have a decrease in some cases are marked high uncertainty. B.) Projected increase in irrigation rate by 2090 under the A1B scenario, ensemble median. Most states have an increase in irrigation rate under all emission scenarios in more than 80% of the GCMs in the ensemble; those that have a decrease in some cases are marked high uncertainty.

This expansion of agricultural area due to climate change will pose an adaptation challenge to some states that have traditionally had a very small percentage of their agricultural land irrigation. One way to see the magnitude of this challenge is to look at the fraction of agricultural water withdrawals in 2090 that would go to fields that were not irrigated in 2005 (Figure 7). In some states with currently limited irrigation systems, more than half of all irrigation withdrawals in 2090 will be for fields that were not irrigated in 2005. Changes in

Kansas (KS), Texas (TX), and Montana (MT) are particularly notable, since these states have a lot of agricultural area that is currently mostly rain-fed but will have significant expansion of their irrigation systems. Conversely, Arizona (AZ) and California (CA) already have a large fraction of their agricultural area irrigated and there is little potential for expansion to increase withdrawals in 2090.

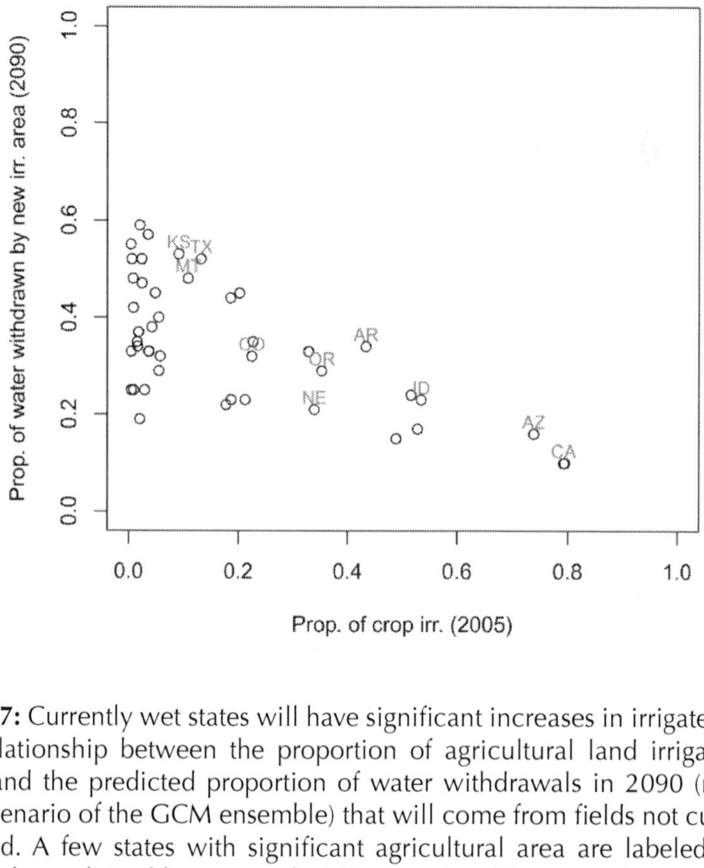

Figure 7: Currently wet states will have significant increases in irrigated area. The relationship between the proportion of agricultural land irrigation in 2005 and the predicted proportion of water withdrawals in 2090 (median A1B scenario of the GCM ensemble) that will come from fields not currently irrigated. A few states with significant agricultural area are labeled, using standard two-digit abbreviation for US states (see Figure 3). Three states are excluded from this graph, because climate change will have a net decrease on irrigated area there (Massachusetts, Connecticut, and Rhode Island).

Installing irrigation in new fields costs money, and producers will only install irrigation if they find the increased revenue from crops grown on irrigated fields worth more than the cost of irrigation. While a full analysis of the costs and benefits of installing irrigation is beyond the scope of the paper, some insights can be gained by looking at state-level averages (Figure 8). In some states that will have a large increase in proportion of agricultural area irrigated, current market value per ha of cropland is fairly high, and may be enough to support the costs of installing irrigation equipment. Conversely, producers in states with lower market value per ha may find it difficult to support the installation of irrigation equipment unless new crops or markets are found. For instance, both Arizona (AZ) and Texas (TX) are projected to have increased irrigated area in response to climate change. However, Arizona has a relatively high average market value for cropland of \$3700 ha^{-1}, driven by production of high-value crops like vegetables and horticultural products, while Texas has a relatively low average market value for cropland of \$420 ha^{-1}, primarily from production of grains and cotton.

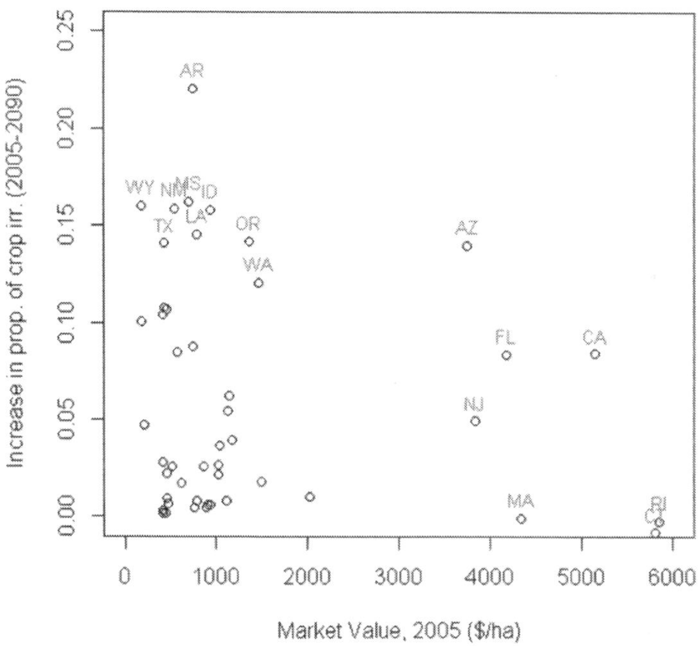

Figure 8: Market value and increase in proportion irrigated. The relationship between the average per hectare market value of cropland in 2005 and the predicted change in proportion of agricultural area irrigated (median A1B scenario of the GCM ensemble). A few states are labeled, using standard two-digit abbreviation for US states (see Figure 3).

Projected Impact of Climate Change on Irrigation Rate

Increases in moisture deficit caused by climate change will likely increase the irrigation rate (Figure 5C). Under all emissions scenarios, irrigation rates will increase with climate change from 7,963 to 8,400–10,415 m^3ha^{-1} in 2090 if the current mix of irrigation technologies stays in place. The B1 emissions scenario has the smallest rate (8,400–9,145 m^3ha^{-1}), while the A2 emissions scenario has the largest rate (9,380–10,415 m^3ha^{-1}). However, if the current trend away from surface irrigation continues at the same pace into the future, needed irrigation rates would decline from current levels to 4,911–6,205 m^3ha^{-1} in 2090 (B1 rate: 4,911–5,355 m^3ha^{-1}, A2 rate: 5,475–6,205 m^3ha^{-1}). Note that our projection only estimates irrigation rates if current observed trends continue, after a moderate degree of adaptation: our methodology implicitly assumes farmers in newly dry climates will irrigate at a rate like that of comparable dry climates. Water policy or water availability might constrain increases in the irrigation rate. Alternatively, failure of farmers to adapt to climate change as much as is assumed in our model would imply an even larger increase in the irrigation rate.

Note also that higher CO_2 levels might increase the water-use efficiency of crop plants, which would further lower the curves in Figure 5C. However, the degree to which this water-use efficiency increase from higher CO_2 concentrations will occur in real systems is unclear, depending on the water balance of the site and the plant species involved [32]–[38]. For one important crop species, wheat, increases in water-use efficiency due to higher CO_2concentrations may exceed 20% in some circumstances [37], [39], [40]. Interestingly, the $\alpha_{t,1}$terms in our model were not statistically significant and are not included in our final model, so there is no clear historical effect of rising ambient CO_2 concentrations on crop water-use efficiency.

The second challenge that climate change poses for irrigation in the US will be this increase in irrigation rates, particularly important in already arid climates. If the current mix of irrigation technologies continues into the future, the biggest increase in irrigation rate is in the Western US (Figure 6B). In places like Arizona, California, Montana, Nevada, and Idaho, farmers will have incentive to apply more water to irrigated lands. However, in some of these states (Arizona, California, and Nevada) there is high uncertainty in the predictions of the ensemble of the GCMs (Figure 6B).

Moving away from surface irrigation could generate substantial irrigation water-use efficiency gains, offsetting demand for an increase in irrigation rate in these states (Figure 9). States with the greatest potential for decrease in irrigation rate with a move away from surface irrigation are Arkansas (AR), Louisiana (LA), Arizona (AZ), Wyoming (WY), and Colorado (CO). In these states, the potential for decreased irrigation rate can more than offset the potential for irrigated area expansion due to climate change. On the other hand, some states such as Iowa and Illinois currently use very little surface irrigation, and there is little potential for decreasing irrigation rates with a move away from surface irrigation. Note also that there is considerable variability among GCMs for the Southwest, and while the median projection under all emissions scenarios is a large increase in irrigation rates in these states, a few GCMs actually project a decrease in moisture deficit and hence a decrease in irrigation rates in these states.

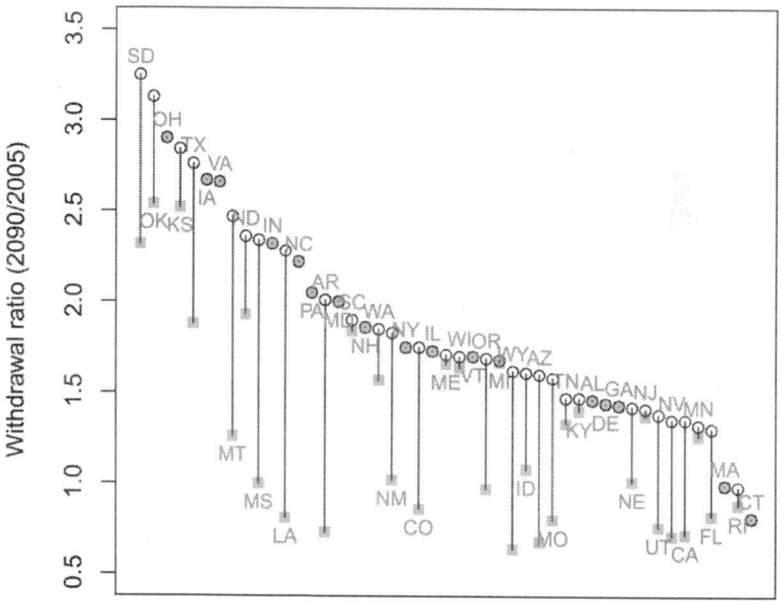

Figure 9: Withdrawal ratios by state. The ratio of irrigation withdrawals in 2090 to irrigation withdrawals in 2005 if the mix of irrigation technologies

stays the same as today (open circles) or if the current trend away from surface irrigation continues into the future (grey squares). A ratio of more than one indicates withdrawals will increase, while a value of less than one indicates withdrawals will decrease. Data points are labeled using the standard two-digit abbreviation for US states (see Figure 3), staggered so that labels do not overlap.

In the absence of gains in irrigation water-use efficiency, total water withdrawals for irrigation are likely to increase substantially (Figure 5D). This is true over all emissions pathways, but the increase in particularly striking for the A2 emissions pathway. Thus, irrigation water-use efficiency gains from switching away from surface irrigation can be an effective adaptation under moderate climate change, but cannot be wholly effective if climate change is severe. Moreover, increases in irrigation efficiency can only serve as adaptation effectively in some states (Figure 9).

DISCUSSION

Our results suggest that there are two different challenges that climate change will pose to agricultural irrigation in the US. In states in currently relatively wet climates, expansion of irrigated area is likely to be an important driver in increases in overall agricultural withdrawals. In some cases, these states have relatively little experience managing water use under scarcity, at least compared to states in the western US. For instance, many states in the eastern United States use a riparian water rights system, where all land owners along a waterbody have a right to make reasonable use of water at any time. In these states, there may currently be little regulation on the installation of a new irrigation system. Our results suggest that a key goal for these states should be developing appropriate regulations that make sure new irrigation systems have access to sufficient water without affect other water users in the area.

In contrast, in states in currently dry climate that already have a large fraction of agricultural area irrigated, our results project the predominant driver of increases in water withdrawal will be the tendency of producers to apply more irrigation water in response to droughts. Fully using technologies to increase irrigation efficiency will be a key goal for these states. The switch away from surface irrigation is included in our model, but many other forms of adaptation are possible[2], [41]. Changes in agricultural practices such as tilling and plant spacing can affect water demand, as can changes in crop varieties beyond what is implicitly assumed by our model. Many of these currently dry states are in the western US and follow a prior appropriation water rights system, where older uses of water have priority over new uses of water. New users of water may be limited by their water right in how much they can increase their irrigation volume. However, currently many states lack mechanisms for water shortages to affect the price or distribution of existing water rights to farmers, leaving them little incentive to consider low water availability in their individual decision-making [9], [42]–[45]. Farmers would have more incentive to switch to more water-efficient crops if the price of irrigation water more accurately reflects its scarcity.

Our results also highlight the importance of greenhouse gas mitigation for reducing impacts on freshwater demand. The median difference in extra demand for withdrawals in the A2 and B1 scenarios is 57 billion m^3 per year in 2090. This means that avoiding the release of 1 GtC will result in the U.S. needing roughly 65 million m^3 less irrigation water annually. Our results demonstrate that, in addition to reducing climate change's many other impacts, climate change mitigation saves irrigation water.

Our results stress that climate change will likely increase irrigation water use in the US, consistent with some previously published research [17], [18], [21], but in contrast with the results of other studies that showed mixed results depending on the GCM examined [19], [20]. At the same time, other research suggests climate change will decrease water supply, at least in some regions, either because of changes in annual supply or because of intra-annual changes in the hydrologic cycle such as changes in snowpack [19]. The exact response of agricultural producers depends on irrigation policy and law, the price of irrigation and water, and the income from crop production, among other things.

The central assumption when making climate change projections with such a statistical model is that past is prologue: how agricultural irrigation changed in response to past climate events is indicative of how it will respond in the future. While this is a reasonable assumption used by many modeling studies, readers should remember that more radical changes in water governance or water availability in the U.S. are theoretically possible, and would imply different forecasts than the ones we present.

Further research into the mechanisms that control the decisions of agricultural producers is key to evaluating the extent to which observed trends in the past two decades are indicative of the response of the agricultural system climate change. For instance, additional field-based studies of the impact of increased CO_2 on crop water use efficiency are needed to incorporate these feedbacks into mechanistic models of responses to climate change. Other factors known to affect irrigation withdrawals include the price of irrigation and income from commodity production.

Our simple statistical model provides a useful complement to more complex, mechanistic models of agricultural irrigation use. This model points out on a national scale how climate change is projected to impact agricultural water demand in different states, and how those states will likely be driven to change their agricultural practices in response to these impacts–either through increasing the area of agricultural land irrigated or through increasing the irrigation rate on already irrigated land. This information is useful to agricultural and water resource managers for informing their long-term planning, and provides insight more broadly into how changes to agricultural water demand will likely complicate water management in the future, among many other demands from other sectors.

ACKNOWLEDGMENTS

RIM is grateful for an appointment as an affiliate of the CONS Program, University of Maryland at College Park, and EHG is grateful for his affiliate appointment in the University of Washington, School of Forest Resources.

REFERENCES

1. IPCC (2007) Climate Change 2007: The Physical Science Basis; Change IPoC, editor. Cambridge, UK: Cambridge University Press.

2. Howden SM, Soussana JF, Tubiello FN, Chhetri N, Dunlop M, et al. (2007) Adapting agriculture to climate change. Proceedings of the National Academy of Sciences of the United States of America 104: 19691–19696. doi: 10.1073/pnas.0701890104
3. Schmidhuber J, Tubiello FN (2007) Global food security under climate change. Proceedings of the National Academy of Sciences of the United States of America 104: 19703–19708. doi: 10.1073/pnas.0701976104
4. Tubiello FN, Fischer GI (2007) Reducing climate change impacts on agriculture: Global and regional effects of mitigation, 2000–2080. Technological Forecasting and Social Change 74: 1030–1056. doi: 10.1016/j.techfore.2006.05.027
5. Fischer G, Shah M, Tubiello FN, van Velhuizen H (2005) Socio-economic and climate change impacts on agriculture: an integrated assessment, 1990–2080. Philosophical Transactions of the Royal Society B-Biological Sciences 360: 2067–2083. doi: 10.1098/rstb.2005.1744
6. Schlenker W, Roberts MJ (2009) Nonlinear temperature effects indicate severe damages to US crop yields under climate change. Proceedings of the National Academy of Sciences of the United States of America 106: 15594–15598. doi: 10.1073/pnas.0906865106
7. Tubiello FN, Soussana JF, Howden SM (2007) Crop and pasture response to climate change. Proceedings of the National Academy of Sciences of the United States of America 104: 19686–19690. doi: 10.1073/pnas.0701728104
8. Luedeling E, Steinmann KP, Zhang MH, Brown PH, Grant J, et al. (2011) Climate change effects on walnut pests in California. Global Change Biology 17: 228–238. doi: 10.1111/j.1365-2486.2010.02227.x
9. Rosegrant MW, Ringler C, Zhu TJ (2009) Water for Agriculture: Maintaining Food Security under Growing Scarcity. Annual Review of Environment and Resources 34: 205–222. doi: 10.1146/annurev.environ.030308.090351
10. Connor J, Schwabe K, King D, Kaczan D, Kirby M (2009) Impacts of climate change on lower Murray irrigation. Australian Journal of Agricultural and Resource Economics 53: 437–456. doi: 10.1111/j.1467-8489.2009.00460.x

11. Taylor R, Scanlon B, Doll P, Rodell M, van Beek R, et al.. (2012) Ground water and climate change. Nature Climate Change 1744: http://dx.doi.org/10.1038/nclimate1744.
12. Sun G, McNulty SG, Myers JAM, Cohen EC (2008) Impacts of Multiple Stresses on Water Demand and Supply Across the Southeastern United States. Journal of the American Water Resources Association 44: 1441–1457. doi: 10.1111/j.1752-1688.2008.00250.x
13. Palmer MA, Liermann CAR, Nilsson C, Florke M, Alcamo J, et al. (2008) Climate change and the world's river basins: anticipating management options. Frontiers in Ecology and the Environment 6: 81–89. doi: 10.1890/060148
14. Alcamo J, Florke M, Marker M (2007) Future long-term changes in global water resources driven by socio-economic and climatic changes. Hydrological Sciences Journal-Journal Des Sciences Hydrologiques 52: 247–275. doi: 10.1623/hysj.52.2.247
15. Elbakidze L (2006) Potential economic impacts of changes in water availability on agriculture in the Truckee and Carson River Basins, Nevada, USA. Journal of the American Water Resources Association 42: 841–849. doi: 10.1111/j.1752-1688.2006.tb04498.x
16. Bureau of Reclamation (2011) SECURE Water Act Section 9503(c)- Reclamation Climate Change and Water, Report to Congress. Washington, DC: US Department of the Interior, Bureau of Reclamation. 206 p.
17. Fischer G, Tubiello FN, Van Velthuizen H, Wiberg DA (2007) Climate change impacts on irrigation water requirements: Effects of mitigation, 1990–2080. Technological Forecasting and Social Change 74: 1083–1107. doi: 10.1016/j.techfore.2006.05.021
18. Schlenker W, Hanemann WM, Fisher AC (2005) Will US agriculture really benefit from global warming? Accounting for irrigation in the hedonic approach. American Economic Review 95: 395–406. doi: 10.1257/0002828053828455
19. Thomson AM, Rosenberg NJ, Izaurralde RC, Brown RA (2005) Climate change impacts for the conterminous USA: An integrated assessment - Part 5. Irrigated agriculture and national grain crop production. Climatic Change 69: 89–105. doi: 10.1007/s10584-005-3611-x

20. Rosenzweig C, Strzepek KM, Major DC, Iglesias A, Yates DN, et al. (2004) Water resources for agriculture in a changing climate: international case studies. Global Environmental Change-Human and Policy Dimensions 14: 345–360. doi: 10.1016/j.gloenvcha.2004.09.003
21. Schlenker W, Hanemann WM, Fisher AC (2007) Water availability, degree days, and the potential impact of climate change on irrigated agriculture in California. Climatic Change 81: 19–38. doi: 10.1007/s10584-005-9008-z
22. Kenny J, Barber N, Hutson S, Linsey K, Lovelace J, et al.. (2009) Estimated Use of Water in the United States in 2005. Reston, VA: U.S. Geological Survey.
23. USDA (2008) Farm and Ranch Irrigation Survey. Washington, DC: U.S. Department of Agriculture.
24. USDA (2009) US Census of Agriculture:2007. Washington, DC: United States Department of Agriculture.
25. Daly C, Neilson RP, Phillips DL (1994) A statistical-topographic model for mapping climatological precipitation over mountainous terrain. Journal of applied Meteorology 33: 140–158. doi: 10.1175/1520-0450(1994)033<0140:astmfm>2.0.co;2
26. Meehl G, Covey C, Delworth T, Latif M, McAvaney B, et al.. (2007) The WCRP CMIP3 Multimodel Dataset: A new era in climate change research. Bulletin of the American Meterological Society September: 1383–1394.
27. Girvetz EH, Zganjar C, Raber GT, Maurer EP, Kareiva P, et al.. (2009) Applied Climate-Change Analysis: The Climate Wizard Tool. Plos One 4.
28. IPCC (2000) Special Report on Emissions Scenarios. Geneva: Intergovernmental Panel on Climate Change.
29. Lu J, Sun G, McNulty SG, Amataya D (2005) A comparison of six potential evapotranspiration methods for regional use in the southeastern United States. Journal of the American Water Resources Association 41: 621–633. doi: 10.1111/j.1752-1688.2005.tb03759.x
30. Vörösmarty CJ, Federer C, Schloss AL (1998) Potential evapotranspiration functions compared on U.S. watersheds: Possible implications for global-scale water balance and terrestrial

ecosystem modelling. Journal of Hydrology 207: 147–169. doi: 10.1016/s0022-1694(98)00109-7

31. Kingston D, Todd M, Taylor R, Thompson JN, Arnell NW (2009) Uncertainty in the estimation of potential evapotranspiration under climate change. Geophysical Research Letters 36: 1–6. doi: 10.1029/2009gl040267
32. Penuelas J, Canadell JG, Ogaya R (2011) Increased water-use efficiency during the 20th century did not translate into enhanced tree growth. Global Ecology and Biogeography 20: 597–608. doi: 10.1111/j.1466-8238.2010.00608.x
33. Huang JG, Bergeron Y, Denneler B, Berninger F, Tardif J (2007) Response of forest trees to increased atmospheric CO_2. Critical Reviews in Plant Sciences 26: 265–283. doi: 10.1080/07352680701626978
34. Morgan JA, Pataki DE, Korner C, Clark H, Del Grosso SJ, et al. (2004) Water relations in grassland and desert ecosystems exposed to elevated atmospheric CO_2. Oecologia 140: 11–25. doi: 10.1007/s00442-004-1550-2
35. Wullschleger SD, Tschaplinski TJ, Norby RJ (2002) Plant water relations at elevated CO_2 - implications for water-limited environments. Plant Cell and Environment 25: 319–331. doi: 10.1046/j.1365-3040.2002.00796.x
36. Medlyn BE, Barton CVM, Broadmeadow MSJ, Ceulemans R, De Angelis P, et al. (2001) Stomatal conductance of forest species after long-term exposure to elevated CO_2 concentration: a synthesis. New Phytologist 149: 247–264. doi: 10.1046/j.1469-8137.2001.00028.x
37. Moore BD, Cheng SH, Sims D, Seemann JR (1999) The biochemical and molecular basis for photosynthetic acclimation to elevated atmospheric CO_2. Plant Cell and Environment 22: 567–582. doi: 10.1046/j.1365-3040.1999.00432.x
38. Drake BG, GonzalezMeler MA, Long SP (1997) More efficient plants: A consequence of rising atmospheric CO_2? Annual Review of Plant Physiology and Plant Molecular Biology 48: 609–639. doi: 10.1146/annurev.arplant.48.1.609
39. Manderscheid R, Weigel HJ (2007) Drought stress effects on wheat are mitigated by atmospheric CO_2 enrichment. Agronomy for Sustainable Development 27: 79–87.

40. Hunsaker DJ, Kimball BA, Pinter PJ, Wall GW, LaMorte RL, et al. (2000) CO_2 enrichment and soil nitrogen effects on wheat evapotranspiration and water use efficiency. Agricultural and Forest Meteorology 104: 85–105. doi: 10.1016/s0168-1923(00)00157-x
41. Easterling WE (1996) Adapting North American agriculture to climate change in review. Agricultural and Forest Meteorology 80: 1–53. doi: 10.1016/0168-1923(95)02315-1
42. Hellegers P, Perry CJ (2006) Can irrigation water use be guided by market forces? Theory and practice. International Journal of Water Resources Development 22: 79–86. doi: 10.1080/07900620500405643
43. Tardieu H, Prefol B (2002) Full cost or "sustainability cost" pricing in irrigated agriculture. Charging for water can be effective, but is it sufficient? Irrigation and Drainage 51: 97–107. doi: 10.1002/ird.44
44. Savenije HHG, van der Zaag P (2002) Water as an economic good and demand management - Paradigms with pitfalls. Water International 27: 98–104. doi: 10.1080/02508060208686982
45. Hooker MA, Alexander WE (1998) Estimating the demand for irrigation water in the Central Valley of California. Journal of the American Water Resources Association 34: 497–505. doi: 10.1111/j.1752-1688.1998.tb00949.x

Chapter 6

Application of Hydrologic Tools and Monitoring to Support Managed Aquifer Recharge Decision Making in the Upper San Pedro River, Arizona, USA

Laurel J. Lacher [1], Dale S. Turner [2], Bruce Gungle [3], Brooke M. Bushman [2] and Holly E. Richter [2]

[1]Lacher Hydrological Consulting, 821 S. Meyer Avenue, Tucson, AZ 85701, USA

[2]The Nature Conservancy, 1510 E. Fort Lowell Road, Tucson, AZ 85719, USA

[3]Arizona Water Science Center—U.S. Geological Survey, 520 N Park Avenue, Tucson, AZ 85719, USA

ABSTRACT

The San Pedro River originates in Sonora, Mexico, and flows north through Arizona, USA, to its confluence with the Gila River. The 92-km Upper San Pedro River is characterized by interrupted perennial flow, and serves as a vital wildlife corridor through this semiarid to arid region. Over the past century, groundwater pumping in this bi-national basin has depleted baseflows in the river. In 2007, the United States Geological Survey published the most recent groundwater model of the basin. This model served as the basis for predictive simulations, including maps of stream flow capture due to pumping and of stream flow restoration due to managed aquifer recharge. Simulation results show that ramping up near-stream recharge, as needed, to compensate for downward pumping-related stress on the water table, could sustain baseflows in the Upper San Pedro River at or above 2003 levels until the year 2100 with less than 4.7 million cubic meters per year (MCM/yr). Wet-dry mapping of the river over a period of 15 years developed a body of empirical evidence which, when combined with the simulation tools, provided powerful technical support to decision makers struggling to manage aquifer recharge to support baseflows in the river while also accommodating the economic needs of the basin.

INTRODUCTION

Social Context

Balancing the social and economic water needs of humans with those of the environment, frequently referred to as sustainable water management, presents a global challenge, especially in arid and semi-arid regions, such as the American Desert Southwest. Understanding that the concept of sustainable water management is subjective, particularly in areas where human groundwater extractions have long exceeded average annual natural recharge, this study focuses on the tension between human demands for groundwater and the groundwater needs of an exceedingly rare perennial river system within the same basin of Southeast Arizona.

We report here on a combination of hydrologic tools and monitoring that are supporting significant progress toward many measures of sustainability on the scale of decades at one important site. The case for developing a regional groundwater recharge network along the Upper San Pedro River is built upon a series of studies and reports funded by the Upper San Pedro Partnership (Partnership), and vetted by its Technical Committee over a period of approximately 15 years. The Partnership is a consortium of 23 agencies and organizations working together to meet the long-term water needs of the Sierra Vista Subwatershed by achieving sustainable yield of the regional aquifer to: (1) preserve the San Pedro Riparian National Conservation Area (SPRNCA); and (2) ensure the long-term viability of Fort Huachuca. The purpose of the Partnership is to coordinate and cooperate in the identification, prioritization and implementation of comprehensive policies and projects to assist in meeting water needs in the Sierra Vista Subwatershed of the Upper San Pedro River Basin. Richter and others [1] provide a more complete discussion of the purpose, methods, and vision encompassed by the Partnership. In addition, recent groundwater modeling efforts have focused stakeholders' collective understanding of the hydrologic system on actionable strategies. Together, this body of work helps identify potential spatial and temporal groundwater recharge targets for mitigating many of the negative environmental, regulatory and economic consequences that may result if groundwater inputs to the San Pedro River diminish as a result of pumping, as predicted by groundwater modeling. Using this information, three of the stakeholders including the U.S. Army Compatible Use Buffer (ACUB) Program, Cochise County, and The Nature Conservancy are collaborating to develop an aquifer protection and recharge network of sites in the San Pedro River basin to protect flows in the river from anticipated pumping-related depletions over the next 50 to 100 years. This interim solution is intended to allow time to develop other longer-term strategies for addressing issues of the accumulated groundwater deficit in the regional aquifer and climate change. Examples of such longer-term strategies could include gradual elimination of consumptive use of groundwater, limitations on new pumping, enhanced utilization of urban runoff, and importation of water from outside the basin. Most of these strategies are controversial and/or expensive, and would require significant political and legal efforts to secure physical water supplies. This approach—using the best available science to implement a relatively uncontroversial and legally available interim solution to preserve baseflows while more difficult long-term solutions are pursued—may serve as a model for other dry-land river basins.

Study Area

The San Pedro River flows north 279 kilometers (km) from its headwaters in northeastern Sonora, Mexico, to its confluence with the Gila River in southeastern Arizona, U.S.A. This paper focuses on the 2460-km^2 Sierra Vista Subwatershed ("subwatershed") of the Upper San Pedro basin ("basin"), north of the United States-Mexico boundary (Figure 1). This subwatershed area includes about 47 km of the river, most of it within the San Pedro Riparian National Conservation Area (SPRNCA), designated by the United States Congress in 1988 to protect and enhance the riparian area and its aquatic resources [2].

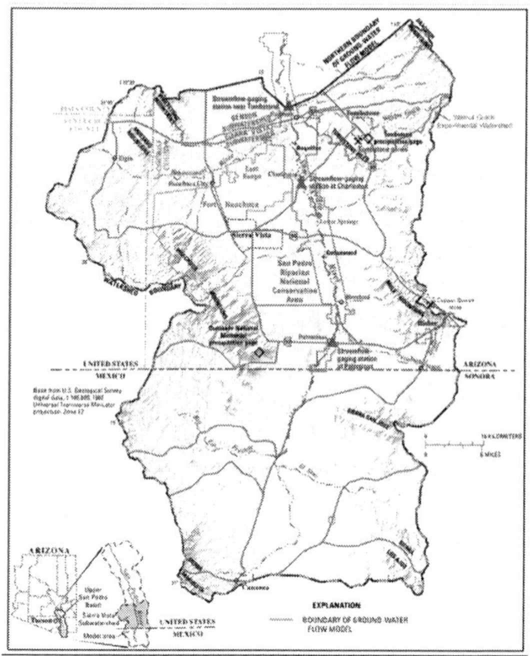

Figure 1. Map of the Upper San Pedro Basin showing the extent of the USGS groundwater flow model [3] and the San Pedro Riparian National Conser-

vation Area. The Sierra Vista subwatershed is the area north of the United States—Mexico boundary.

Elevations within the Sierra Vista subwatershed range from more than 2800 meters (m) above mean sea level in the Huachuca Mountains on the western edge of the basin, to 1052 m at the Tombstone stream-flow gaging station at the north (downstream) end of the subwatershed. Precipitation across the topographically diverse basin ranges from 35 centimeters per year (cm/yr) near Tombstone in the central-north part of the subwatershed to about 76 cm/year in the highest parts of the Huachuca Mountains [4,5]. Like other areas in the desert southwest, precipitation in the study area is predictably bimodal with a summer monsoon season from July through mid-September accounting for about one half, and a winter wet season from December through March that accounts for another one third of the annual precipitation. Occasionally, tropical storms will trigger very large runoff events in October, but otherwise the fall months are typically dry. Stream flow is lowest in May and June, with peak monthly flows occurring during the summer monsoon season (Figure 2). Average monthly minimum flows range from about 1300 cubic meters per day (cu-m/d) at the Palominas stream-flow gaging station in the south (upstream) end of the subwatershed to 19,300 cu-m/d at the Charleston station, and 8100 cu-m/d at the Tombstone station on the north (downstream) end of the study area (Figure 2). Snow melt in the Huachuca Mountains often supports flow in mountain springs and intermittent streams in the spring months of March and April.

Groundwater discharge from the regional basin-fill aquifer supports baseflows in the river and evapotranspiration by near-stream phreatophytes. Groundwater pumping from that same aquifer supplies a human population in the region that is expected to increase by 46% by the year 2050 [6]. Sierra Vista, Arizona, is the largest municipality in the basin with a population of roughly 45,300 [7]. The U.S. Department of Defense installation at Fort Huachuca borders Sierra Vista on the west and north sides and is a major economic force in the basin [8].

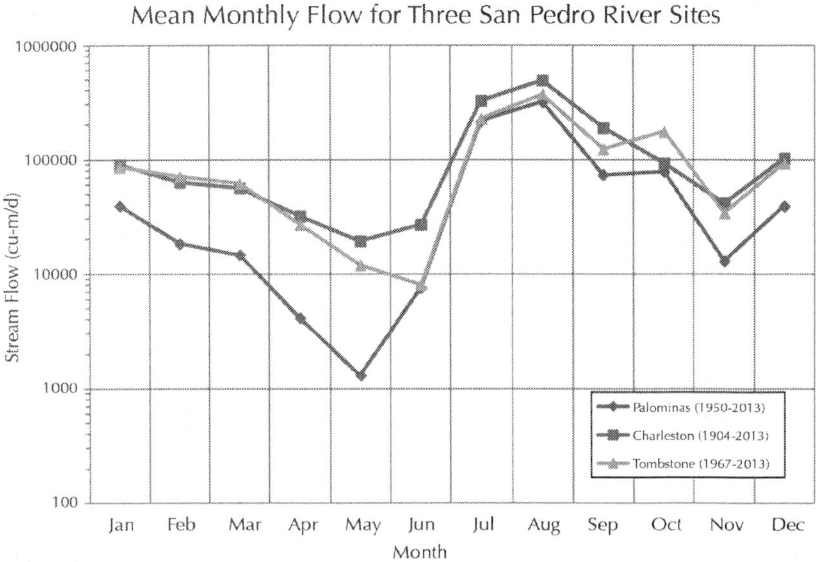

Figure 2. Mean monthly stream flow for the three long-term monitoring sites (shown in Figure 1) on the Upper San Pedro River in the Sierra Vista subwatershed for the period of record at each site. The y-axis is plotted on a log scale to highlight the minimum flows in May and June.

Because of its federal designation and its globally significant biodiversity [9], the SPRNCA has been the subject of extensive hydrological and ecological research. One major finding was that stream flow permanence explained most variance in the basal area of two keystone riparian species, Fremont cottonwood (*Populus fremontii*) and Goodding willow (*Salix gooddingii*) [10,11]. As a result, conservation strategies have focused on maintaining or increasing stream flow through reductions in groundwater pumping and managed aquifer recharge at key locations.

Hydrological Studies in the Basin

General Hydrologic Processes

The Upper San Pedro River is one of the most intensively studied semi-arid stream systems in the world. Estimated ages of the earliest paleo-Indian sites along the banks of the ancestral San Pedro River date back to 11,000 to 13,000 years ago [12,13]. Two Clovis Indian sites have been associated with mammoth kills in the subwatershed, providing some of the earliest dates for human hearths of this culture [13]. The basin has been grazed since the time of the first Spanish explorers in the 1500s [14], and has hosted two major metals mining operations in Tombstone (late 1800s) and Bisbee (1880s–1975), Arizona. In addition, Cananea, Mexico boasts a major copper mine that is currently in active production. The major U.S. Department of Defense Army installation at Fort Huachuca has been in active operation since 1877 and has one of the earliest water rights claims in the basin [15]. More recently, the town of Sierra Vista has grown to include over 45,000 people (including Fort Huachuca). All of these activities have had significant impacts on the landscape and water resources in the basin. Hereford [14] presents a clear description of the land use changes coupled with exceptional periods of flooding near 1900 that led to significant entrenchment and stabilization of the San Pedro River stream channel and dewatering of much of the shallow alluvium of the river channel.

The Walnut Gulch Experimental Station (Figure 1) was established in the northeast area of the subwatershed by the Research Division of the Soil Conservation Service in 1951, and has been the source of continuous instrument-based hydrologic and rangeland studies since that time [16]. In 1966, Brown and others [17] evaluated the water resources of Fort Huachuca, which lies immediately west and north of the City of Sierra Vista (Figure 1). Brown and Aldridge [18] estimated San Pedro River surface discharge from the international boundary with Mexico to its confluence with the Gila River and inputs to the system from tributary inflow and from mountain-front recharge. Much of the assessment of hydrologic resources in the basin that followed came as a byproduct of the development of groundwater models, including those by Freethey [19], Vionnet and Maddock [20], Corell and others [21], Goode and Maddock [22], and Pool and Dickinson [3].

The Arizona Department of Water Resources (ADWR) evaluated the groundwater resources of the basin in 1990 [23], and then again in 2005 [24], in order to determine whether or not the basin should be classified as an Active Management Area (AMA) under the 1980 Groundwater Management Act of Arizona. The ADWR determined that the basin did not meet the statutory criteria for AMA designation [23,24], but much of the research that was conducted as part of that assessment remains highly relevant. Pool and Coes [25] described the state of the knowledge of hydrogeology of the subwatershed. As part of that work, an extensive monitoring program was initiated in the subwatershed that included geophysical surveys, ephemeral stream flow monitoring, installation or refurbishment of a number of stream gaging stations, aquifer storage monitoring using microgravity techniques, and basin-wide water level monitoring. Much of this monitoring program continues today.

In support of the conservation mission of the SPRNCA, Section 321 of the Defense Authorization Act of 2004 [26] revised how the federal Endangered Species Act applies to the Fort Huachuca Military Reservation and directed the Partnership to "...restore and maintain the sustainable yield of the regional aquifer [of the Sierra Vista Subwatershed] by and after September 30, 2011." It also required annual progress reports on these efforts, which were produced for most calendar years from 2002 through 2011 (see [27], for example).

The Partnership sponsored several research reports during this time, which provided the scientific basis for development of the U.S. Geological Survey (USGS) groundwater model by Pool and Dickinson [3]. These studies include Coes and Pool's [28] assessment of ephemeral-stream channel and basin-floor infiltration and recharge, Gungle's [29] analysis of the timing and duration of ephemeral stream flow in the subwatershed, and Leenhouts and others' [30] analysis of the hydrology, vegetation-hydrologic relationships, and evapotranspiration requirements and plant-water sources in the SPRNCA. Leenhouts and others [30] established additional streamflow and groundwater monitoring in the subwatershed, including streamflow stage and permanence data, near-stream alluvial aquifer groundwater and vertical gradient monitoring, and a continuous meteorological and eddy covariance monitoring station for measurement of evapotranspiration in the SPRNCA.

A statistical analysis of the trends in streamflow in the San Pedro River was also published in 2006 by Thomas and Pool [5]. Leake, Pool and Leenhouts [31] used Pool and Dickinson's [3] five-layer groundwater model of the basin to conduct a capture and recharge analysis that mapped the effects of pumping and recharge across the subwatershed on groundwater discharge to the SPRNCA. Kennedy and Gungle [32] analyzed baseflow discharge from the subwatershed at the USGS gaging station near Tombstone, Arizona, at the north end of the subwatershed. Most recently, Lacher [33] updated the Pool and Dickinson [3] groundwater flow model to include recent changes in pumping and artificial recharge in the subwatershed, and simulated the projected effects of population growth-driven increases in pumping on groundwater levels and baseflow through 2105.

Managed Aquifer Recharge

In 2006, Stantec [34] developed a Flood Control Urban Runoff Plan for Cochise County that evaluated the size, placement, and efficacy of 30 existing and proposed stormwater detention basins on the west side of the subwatershed functioning as de facto recharge basins. As part of this study, GeoSystems Analysis [35] used a detailed precipitation-recharge-stormwater runoff regression model, based on the U.S. Department of Agriculture's Automated Geospatial Watershed Assessment Tool [36] for the urbanized Coyote Wash watershed in Sierra Vista to estimate the cost, recharge volume, and urban-enhanced runoff for that watershed and several others on the west side of the San Pedro River within the subwatershed. GeoSystems Analysis previously developed the stormwater regression model used for the Upper San Pedro Partnership and Cochise County based on detailed AGWA stormwater modeling results for the Coyote Wash watershed located in the City of Sierra Vista [37].

Predictive groundwater modeling [33] indicates that within the next 100 years, two regional cones of depression will merge and reduce groundwater flow from the regional aquifer to the San Pedro and Babocomari (a major tributary to the San Pedro) rivers (Figure 3). These simulations incorporate pumping increases over time that reflect U.S. Census growth rates [5,6] and maintain constant natural recharge and evapotranspiration at 2003 levels (as published in the USGS groundwater model [2]) for the entire 21st century. Figure 4a maps the west-east transect A-A′ for the simulated groundwater-level profiles shown in Figure 4b. This transect goes through the Sierra Vista-Fort Huachuca cone of depression, and illustrates how simulated groundwater levels at the groundwater divide between the cone of depression and the river have already declined by 17% since pre-development conditions in 1902 and are predicted to double that level of decline, to 35%, by the year 2100. These reductions in groundwater levels produce commensurate percent reductions in groundwater gradient (change in head divided by distance) as measured from the same point on the groundwater divide to the river. While gradients directly under the river are not yet showing the same degree of impacts as those farther west, the ultimate outcome of these simulated changes would be greater baseflow losses in some losing reaches, smaller gains in some gaining reaches, and possibly the conversion of some reaches from gaining to losing. The resulting groundwater-driven reduction in baseflow and shallow groundwater would likely impair the dependent riparian systems [33].

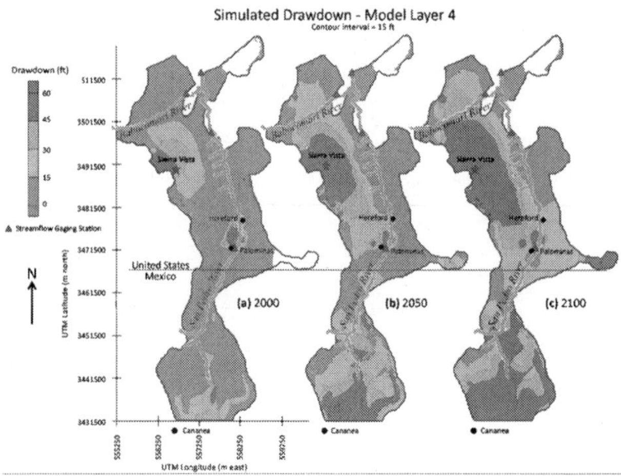

Figure 3. Simulated drawdown (m) in the primary regional aquifer of the basin (a) in 2000; (b) in 2050; and (c) in 2100 [33].

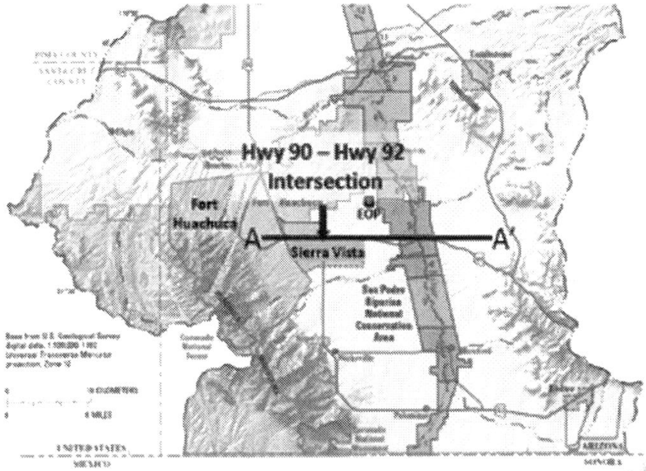

(a)

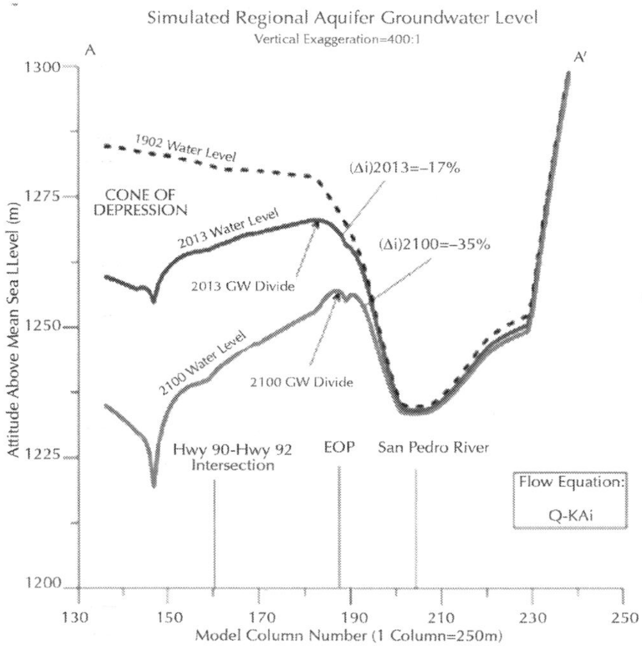

(b)

Figure 4. (a) Cross-section A-A′ in the center of the Sierra Vista subwatershed (adapted from Stromberg and others [10]; (b) Simulated groundwater-level profiles in 1902, 2013, and 2100 along the A-A′ cross-section showing change in gradient (Δi) from −17% in 2013 to −35% in 2100.

One potential solution to the problem of aquifer storage depletion would entail importation of at least some water from a source outside the basin [38]. However, at this time, importation does not appear likely in the foreseeable future (see companion paper by Richter and others [1]). In lieu of a replacement water supply, water managers and stakeholders in the subwatershed recognize the need for a near- to medium-term (*i.e.*, years to decades) intervention to protect stream baseflows and the associated riparian areas.

The City of Sierra Vista has more than 10 years of monitoring data from its Environmental Operations Park (EOP) where treated effluent is recharged at a rate of roughly 9100 cubic-meters per day (m^3/d) through artificial wetlands and recharge basins between the City's pumping center and the San Pedro River. This project was designed and constructed to create a groundwater mound to sustain surface water flow and supplement alluvial groundwater levels during low-flow periods [39]. Groundwater modeling that incorporates the EOP recharge data indicates that the project is successfully contributing recharge to both the nearby cone of depression in the regional aquifer and to baseflows in the San Pedro River [33,40]. Further groundwater modeling presented in this study suggests that additional recharge projects designed to utilize enhanced urban runoff, treated effluent or other local supplies near the San Pedro and Babocomari rivers may successfully mitigate anticipated pumping-related baseflow depletions for up to 100 years.

While conservation efforts within the subwatershed have reduced consumptive use of groundwater in the basin, particularly on the Fort Huachuca Army installation, the cumulative storage loss in the regional aquifer within the subwatershed from the past half century of pumping, estimated to exceed 800 MCM (refer to explanation provided in Section 2.3.2) remains a threat to the San Pedro River and its tributaries by intercepting mountain-front recharge and by reducing the groundwater gradient that drives aquifer discharge to the rivers. Because of this storage deficit and the fact that virtually all water use in the subwatershed depends on groundwater, almost any level of continued groundwater use poses an increasing risk of stream and riparian evapotranspiration capture over time.

Streamflow Permanence Monitoring

Direct measurement of baseflow, defined as the portion of streamflow derived from groundwater, is complicated by the effects of prolonged storm and/or snowmelt runoff (which tend to exaggerate baseflow estimates) and evapotranspiration (which reduces apparent baseflow). Acknowledging these complexities, practitioners have found that inter-annual stream-flow permanence in a river system that experiences intermittent drying in some reaches is a useful indicator of both hydrologic and ecological conditions in the San Pedro basin. Using a technique called wet/dry mapping, analysis of surface water presence (strict criteria are used to ensure consistent definition of surface water versus minimal puddles) during the driest time of year (mid June in the San Pedro Basin) reveals areas with high year-to-year variation in the length of surface wetting (Figure 5). By limiting the influence of storm events as much as possible, and assuming no significant changes in the condition of the riparian forest, these variations in wetted length are believed to be the best available physical expression of local groundwater conditions and may provide early warning of ecological changes [41].

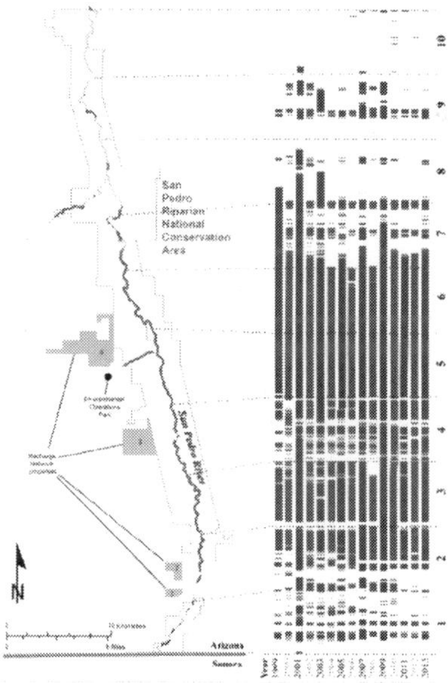

Figure 5. Wet/dry results from the San Pedro Riparian National Conservation Area. The heavy river line shows reaches, which were wet in June 2013.

Bars on right side represent wet reaches for each year, 1999–2013. Labels on the far right identify the 10 analysis segments, each covering 8.1 km. The four green properties (parcels 1 through 4) on map were recently acquired for groundwater protection and recharge projects. Redrawn from Turner and Richter [41].

Wet/dry mapping, as applied in the San Pedro River basin, uses citizen scientists annually to map the spatial extent of surface water in a river or stream. The method provides a comprehensive snapshot of conditions for the whole river at the same date each year, allowing comparisons of year-to-year variability [41]. Beginning in 1999, staff from The Nature Conservancy and the U.S. Bureau of Land Management coordinated volunteers to map the spatial extent of surface water within the SPRNCA. The exact dates varied slightly, but mapping was conducted during the third weekend of June each year to coincide with the lowest flow before the expected start of the summer rainy season. Through the years, progressively more of the river and its tributaries have been surveyed, increasing to 231 km of the mainstem and 266 km of tributaries in 2013. This paper addresses only the 80 km of mainstem river surveyed through SPRNCA.

METHODS

Wet/Dry Mapping

Surveyors walked predetermined segments of the river, recording the coordinates of beginning and end points of all surface water segments greater than or equal to 9.1 m long using paper data forms and consumer-grade Global Positioning System units. They disregarded any dry gaps less than 9.1 m long in otherwise wet reaches. The resulting point coordinates were imported to a Geographic Information System (ArcGIS, Environmental Systems Research Institute, Redlands, CA, USA), and snapped to the closest points on a linear representation of the river. To identify localized trends, organizers partitioned the SPRNCA into 10 equal segments, 8.1 km long. Segments were analyzed for probability of trend using the Mann-Kendall test. For graphic purposes, we calculated and display the Sen estimate of linear trend (detailed methods are provided by Turner and Richter [41]). Starting in 2007, maps and summary data from the wet/dry surveys have been posted each year to a web site (http://www.azconservation.org) for public distribution.

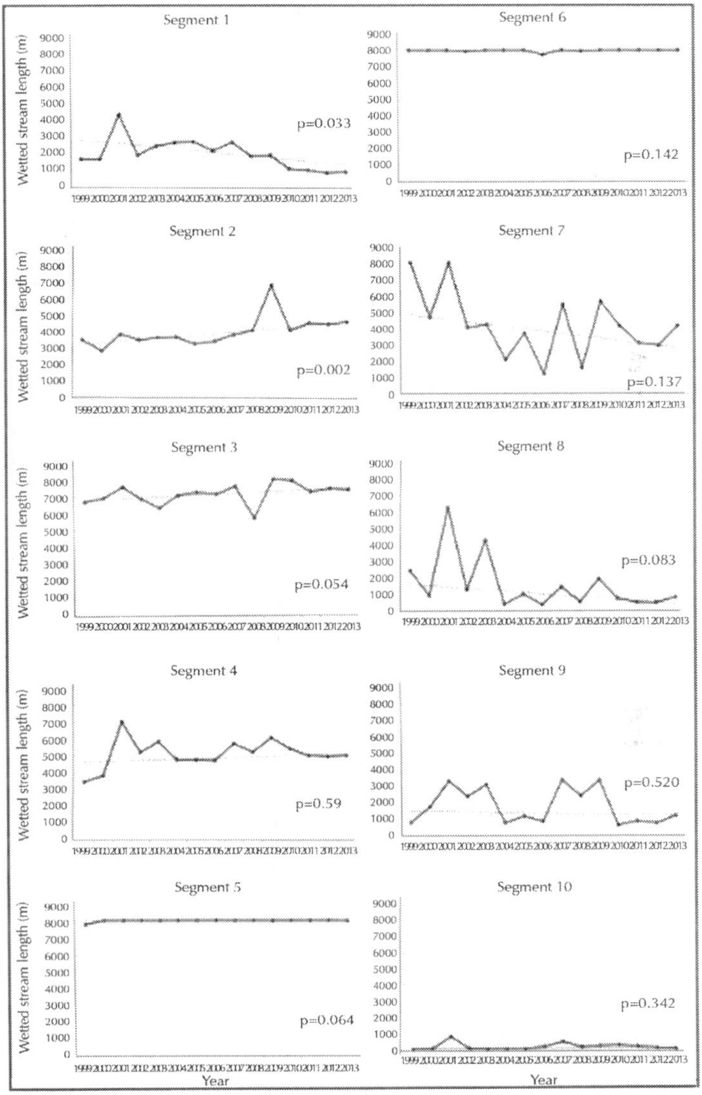

Figure 6. Total wetted lengths for the 10 analysis segments in Figure 3. Segment numbers increase from south to north (downstream). Revised from Turner and Richter [41].

Results from wet/dry mapping in the SPRNCA (Figure 6), as a whole, show about half the river has permanent surface water with some year-to-year variation but no trends. However, analysis of the smaller segments shows considerable variation Turner and Richter [41]). The southernmost segment, Segment 1, displays a significant downward trend, while Segment 2 trends upward. Most of the other segments have either year-to-year up/down results or a stable condition without trend. The distribution of wet and dry reaches has provided a simple way to prioritize and site conservation actions aimed at reduced groundwater pumping and managed aquifer recharge, as described in Section 3 below. Wet/dry mapping data are also expected to provide a quantitative measure of conservation progress after those strategies are implemented.

CAPTURE MAPPING

Overview of Capture

The Upper San Pedro River's position near the geographic center of an alluvial basin makes it a classic example of basin-fill hydrology. The existence of a major pumping center at Sierra Vista/Fort Huachuca between the primary area of recharge for the subwatershed (the Huachuca Mountains on the western boundary) and the river (Figure 1) also makes the basin a classic case study in groundwater capture. Capture is the increase in recharge to, and (or) decrease in discharge from, a basin that eventually occurs as a result of groundwater pumping. Theis [42] first addressed the consequences of groundwater pumping from a previously undeveloped system in 1940:

> "Under natural conditions...previous to development by wells, aquifers are in a state of approximate dynamic equilibrium. Discharge by wells is thus a new discharge superimposed upon a previously stable system, and it must be balanced by an increase in the recharge of the aquifer, or by a decrease in the old natural discharge, or by loss of storage in the aquifer, or by a combination of these."

Lohman and others [43] further clarified the definition of capture from Theis' description:

"The decrease in discharge plus the increase in recharge is termed capture."

This definition of capture may be written as:

$$\text{Capture} = \Delta R + \Delta D \quad (1)$$

where $\boldsymbol{\Delta} R$ and $\boldsymbol{\Delta} D$ equal the increase in recharge and the decrease in discharge, respectively.

In the Upper San Pedro basin, as in many basins in the southwest deserts of the United States, aquifer storage is plentiful as a result of recharge over thousands of years of depositional history, but potential sources of capture are limited primarily to reductions in riparian ET, reductions of groundwater discharge to streams (baseflow), and direct capture of streamflow. The numerical partitioning of extracted groundwater into its constituent sources through a water-budget process is a common and fraught practice in many water-short areas of the United States. Bredehoft, Papodopulos, and Cooper [44] attributed this practice to "perhaps the most common misconception in groundwater hydrology" which is "that a water budget of an area determines the magnitude of possible groundwater development." Under this line of reasoning, many water managers have concluded that all water entering the system as natural recharge is available for extraction without long-term deleterious effects. Brown [45] addressed the problem with this argument through an example of a well whose cone of depression eventually expands to intersect a stream in which he demonstrated that,

"...the rate at which the hydrologic system reaches a new steady state depends on the rate at which the natural discharge [in his example to a stream] can be captured by the cone of depression."

Figure 7 illustrates the evolution of a groundwater system which receives natural recharge (through precipitation) at a fixed and limited rate, R_0, and discharges to a stream as baseflow at a rate of D_0. In Figure 7a, the system is in equilibrium prior to any significant groundwater development. In this natural state of equilibrium (roughly prior to 1940 in the Upper San Pedro basin [3]) recharge equals discharge (Equation (2)):

$$R_{0} = D_{0} \quad (2)$$

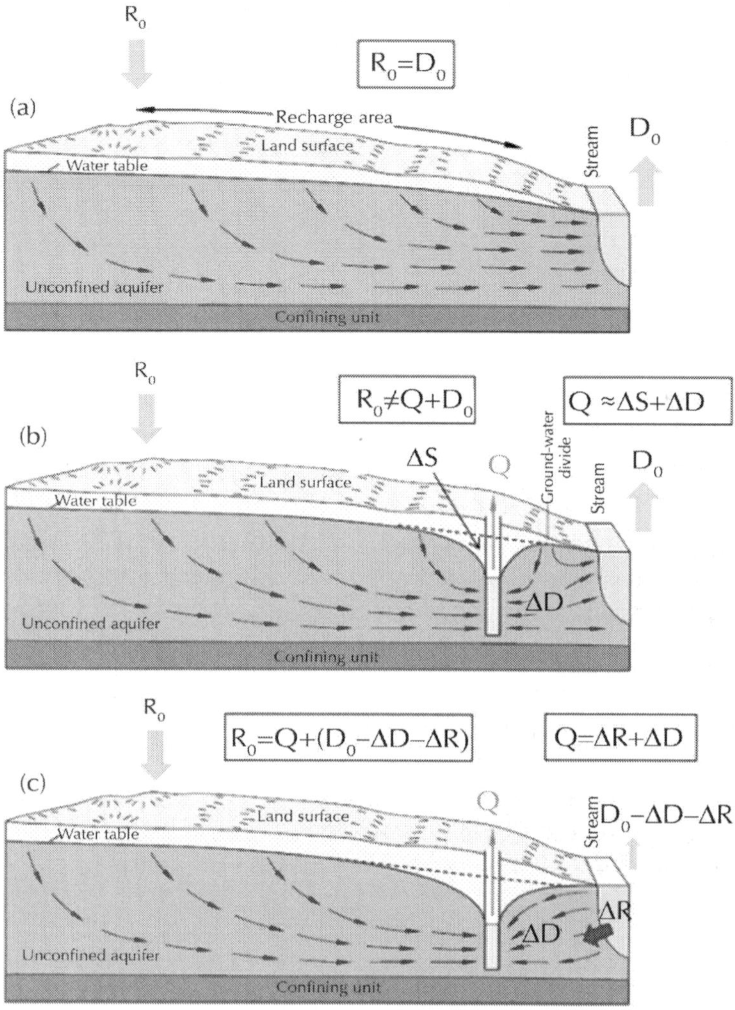

Figure 7. (a) Groundwater discharging to stream under equilibrium conditions; (b) Pumping at rate Q from a well intercepting groundwater that would have discharged to the stream under equilibrium conditions; Q is roughly equal to the rate of change in aquifer storage (ΔS) plus reduced groundwater discharge to the stream (ΔS); (c) Pumping at the same rate (Q) under a new equilibrium condition. Now, pumping is reducing groundwater discharge to the stream (ΔD) and inducing recharge directly from the stream (ΔR), and aquifer storage depletion ceases (ΔS = 0). Adapted from Winter and others ([46], Box C, p. 15) and from Heath ([47], p. 33).

In Figure 7b, pumping at a rate of Q is superimposed on the system, producing a rapid rate of change in aquifer storage near the well (**ΔS**) and intercepting some groundwater that would otherwise have discharged to the river (**ΔD**), but having no immediate effect on discharge to the stream (D_0) due to the persistence of a groundwater "high" between the well and the river. At this point, pumping is roughly equivalent to the rate of change in aquifer storage plus the reduction in groundwater discharge to the stream (Equation (3)).

$$Q=\Delta S+\Delta D \quad (3)$$

and recharge and discharge are no longer in balance (Equation (4)):

$$R_{0\neq Q+ D0} \quad (4)$$

After some time (Figure 7c), the cone of depression intercepts the stream and reverses the gradient of the groundwater so that it now flows toward the well from all directions, directly capturing streamflow. Even though the pumping rate (Q) remains constant, recharge from precipitation also remains fixed at R_0, so the only sources of water for extraction come reduced groundwater discharge to the stream, capture (increased recharge) from the stream, itself, and possibly some additional aquifer storage. A new equilibrium is achieved when aquifer storage is no longer being depleted (**ΔS** $= 0$) and pumping is balanced by the capture of stream flow and decreased discharge to the stream (Equation (5)):

$$R_{0=Q+(D0-\Delta}D-\Delta R) \quad (5)$$

or,

$$Q=\Delta R+\Delta D \quad (6)$$

Lohman [48] identified another potential source of capture not expressly described above. He referred to it as "salvaged rejected recharge from precipitation." This potential source of capture would comprise a new source of recharge (ΔR) as described in Equations (1) and (6). In the case of a large alluvial basin like the San Pedro, where much of the shallow alluvial aquifer is separated by a thick sequence of confining materials from the underlying regional aquifer, this source of potential capture would derive from the occasional replenishment of the shallow alluvial aquifer during runoff from one or more very large precipitation events. Runoff exceeding 25M cu-m/d occurs roughly every 5 to 6 years at the Palominas station (Figure 1), so any excess alluvial aquifer storage opened up by groundwater pumping could capture flood flows that would otherwise have been rejected and remained in the stream.

For the period 2000 to 2009, Kennedy and Gungle [32] found that alluvial aquifer storage changes that govern baseflow measured at the Tombstone stream-flow gaging station (near the far downstream end of the study area (Figure 1)) were a function of upstream riparian ET and summer precipitation. Although they did not identify groundwater pumping as a clear source of baseflow decline in that reach over that short period, they do caution that, "[c]ontinued regional groundwater pumping will, however, eventually lead to a decline in the contribution of regional groundwater to base flow." This contrast between the interannual scale of fluctuations in alluvial aquifer storage and the multi-decadal scale of changes in regional aquifer discharge is highly significant for long-term water management planning. Our study focuses on these longer-term regional aquifer changes, but acknowledges the importance in improving our understanding of the linkages between sources of capture at different temporal and spatial scales.

Groundwater simulations [3,33] indicate that the Upper San Pedro basin is transitioning between the second and third scenarios illustrated in Figure 7. Of course, streamflow is not the only potential source of capture in most groundwater basins, and may not be available at all, as in basins where the streambed has lost connectivity with the underlying aquifer. Other potential sources of capture include riparian evapotranspiration, groundwater inflow from boundary sources, and groundwater outflow from the basin. Because groundwater and surface water systems operate on such different time scales, stresses on a groundwater system may not manifest as baseflow depletions for many years. Unfortunately, "in many circumstances the dynamics of the groundwater system are such that long periods of time are necessary before any kind of an equilibrium condition can develop. In some circumstances the system response is so slow that [groundwater] mining will continue well beyond any reasonable planning period" ([44], p. 55–57). The goal of the study described in this paper is to anticipate the future impacts of 20th and 21st century pumping on the San Pedro River and to develop a strategy to mitigate those effects as they occur.

Development of Capture Map

Leake and others [31,48] developed a unique tool that utilizes the Pool and Dickinson [3] regional groundwater flow model to assess simulated pumping-induced capture of streamflow, spring discharge, and riparian ET across the model domain in response to a unit pumping stress at every location in the model, and presents the results as a map overlay with colored contours representing the amount of capture, as a fraction of the pumping rate, after pumping for a specified amount of time (Figure 8). Leake and others [49] provide detailed steps for developing a capture map using a groundwater model. Alternatively, capture for a given location can be calculated as the pumping time needed to reach a depletion-dominated supply (the time at which capture begins to provide greater than 50% of total groundwater pumped). A typical time scale for a basin in the American Southwest might be 0 to 100 years [50]. Capture map development can also be run in reverse, providing estimates of the total increase in streamflow, riparian evapotranspiration, and spring flow, as a fraction of recharge rate, after recharging at a unit rate for a specified amount of time [51]. The two are not necessarily the inverse of each other. Leake and others [31] calculated stream and riparian area capture resulting from pumping in the lower basin fill primary aquifer (layer 4 of the 5-layer groundwater model [3]) in the Upper San Pedro basin. They used the same method to calculate response in the stream/riparian area from recharge applied to the top-most layer of the model that overlies the extent of layer 4 (layers 1, 2, or 4).

Use of Capture Map in the San Pedro Basin

Richter and others [1] discuss the evolution of the capture map concept and its use in policy development within the basin. While the capture map does not replace groundwater modeling, it is a simple, intuitive tool that permits the layman to gain a better understanding of the degree of connectivity between the groundwater system and the river. Policy makers and stakeholders embraced the capture map in the San Pedro basin as a guide for various preliminary decisions on community development and ecological restoration.

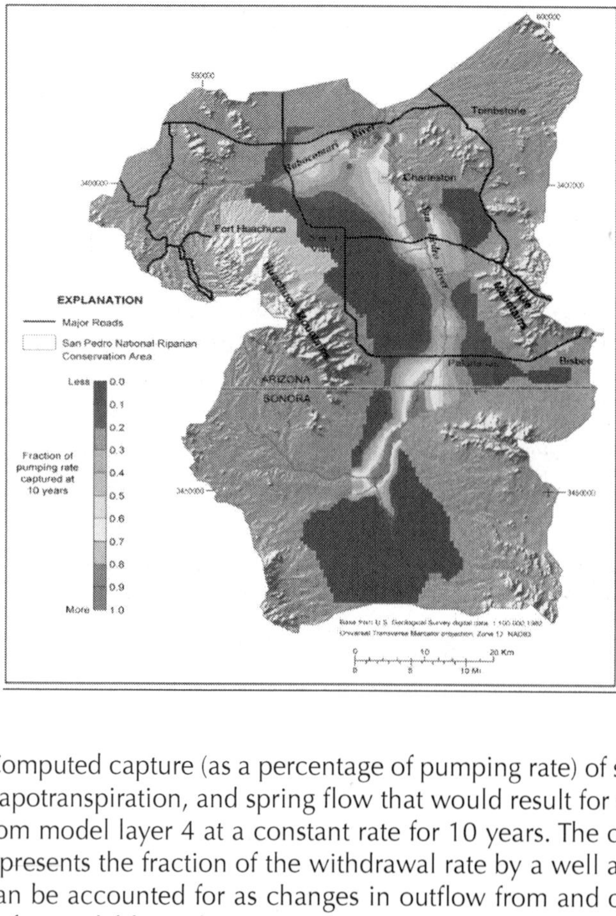

Figure 8. Computed capture (as a percentage of pumping rate) of streamflow, riparian evapotranspiration, and spring flow that would result for withdrawal of water from model layer 4 at a constant rate for 10 years. The color at any location represents the fraction of the withdrawal rate by a well at that location that can be accounted for as changes in outflow from and or inflow to the aquifer for model boundaries representing streams, riparian vegetation, and springs. Redrawn from Leake and others [31].

Advantages/Disadvantages and Limitations of the Tool

Capture maps highlight where pumping will have the greatest impact on a water resource (such as a river) within a specified period of time, and thus make an excellent preliminary planning tool for water managers, commercial and residential developers, and conservationists who can benefit from reducing the immediate impacts of groundwater pumping. The similarly constructed recharge map provides similar planning benefits, offering the planner an initial overview of the most advantageous sites for storm-water, treated effluent, or other recharge facilities.

Despite the value of capture maps for communicating some complex hydrologic concepts to lay audiences, details about specific volumes and rates of capture for specific sources/sinks in the basin cannot be extrapolated directly from this tool. Capture maps, based on linear superposition, reflect the assumption of constant hydrologic properties of the aquifer, assuming that pumping causes no non-linear behavior in the hydrologic system. This limitation, and the fact that each cell of the model is stressed in isolation to make the map, means that capture maps cannot replace the use of a full groundwater model for examining the cumulative impacts of various stresses and sinks that change over time or that cause nonlinearities in hydrologic properties or boundary conditions.

NEAR-STREAM RECHARGE SIMULATIONS

Description of Model

Simulations in this study used the most recent and most sophisticated groundwater flow model of the Upper San Pedro Basin available [3]. Although the model area includes all of the 4500-km^2 basin of which 40% is in Mexico, this study focuses on the Sierra Vista subwatershed within the United States (Figure 1). This MODFLOW-2000 [52] model is based on a uniform 250 m × 250 m grid spacing oriented north-south in alignment with the basin. The stream is represented by the Stream Package [53], hydraulic flow is modeled with the Layer Property Flow (LPF) package, and riparian evapotranspiration is modeled with the Evapotranspiration (EVT) Package of MODFLOW-2000. The two-season model reflects the seasonal significance of evapotranspiration in the riparian zones along the San Pedro and its tributaries. The cool season extends from mid-October to mid-March, and the warm season runs from mid-March through mid-October [3].

As with all groundwater models, this model has several limitations. Most significantly, it does not simulate flood flows, which are known to contribute significant recharge to the alluvial aquifer near the center of the basin as well as along some ephemeral tributaries to the mainstem of the San Pedro River. This seasonal "topping off" of the shallow alluvial aquifer may support baseflows in the river through one or more dry seasons, and is an important component of the riparian system. While this shortcoming means that the model tends to underestimate true baseflow (regional aquifer plus shallow alluvial aquifer contributions), it does not preclude the model as a useful tool for analyzing the effects of pumping on the regional aquifer's contribution to baseflow. For this reason, the term "baseflow" in this study refers strictly to that component of total baseflow that derives from the regional aquifer where most of the pumping in the basin occurs.

Figure 9 illustrates the conceptualized hydrogeologic cross section (upstream view) of the subwatershed and shows how the model layers correspond to that conceptualization. The model structure includes five layers in a stacked-bowl configuration representing sediment accumulation in the structural depression between two bounding mountain ranges, as is typical of the Basin and Range province of the western United States [3]. Only model layer 5 is found throughout the entire region (Figure 10).

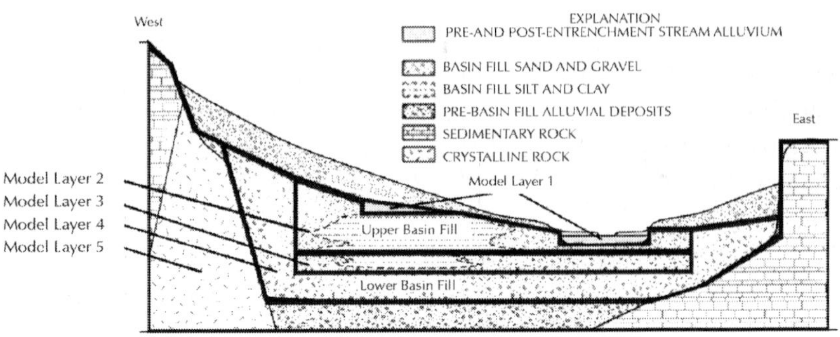

Figure 9. Conceptualized cross section of basin showing model layers. Adapted from Figure 3 in Pool and Dickinson [3]).

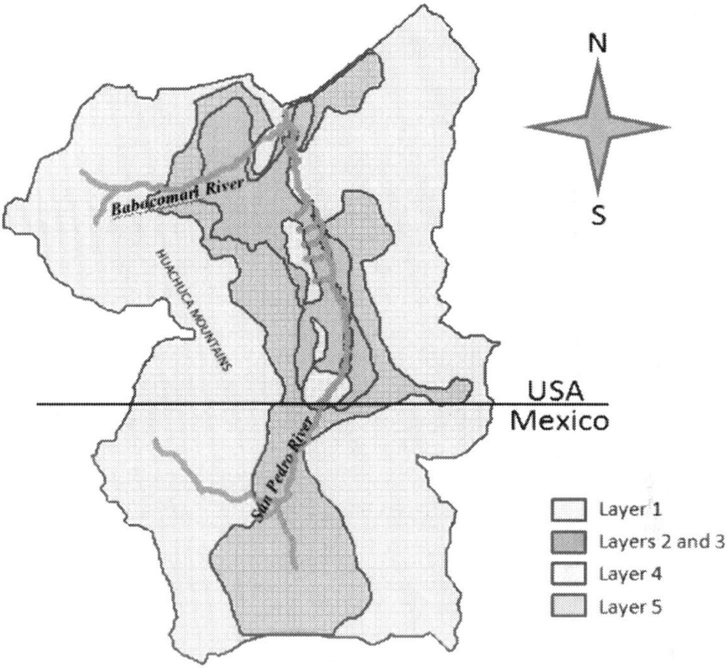

Figure 10. Model layers [2] in plan view with San Pedro River intersecting various layers. Light blue line represents river location but does not necessarily signify perennial flow.

Use of Model in Water Resources Assessments

Initial groundwater modeling efforts by Pool and Dickinson [3] simulated transient groundwater levels and baseflows in the basin resulting from 20th-century (1902–2003) pumping and changes in riparian evapotranspiration associated with climate and evolving geomorphology of the stream channel [14]. These simulations reflected the development of a major cone of depression under the Sierra Vista/ Fort Huachuca population center on the west side of the river, and a smaller area of groundwater depletion very close to the river near the communities of Palominas and Hereford, where agricultural pumping occurred for many decades (Figure 3a). The effects of groundwater use in Mexico also manifest as an irregular cone of depression progressing northward from the south edge of the regional aquifer.

Simulations by Lacher [33] built on the work of Pool and Dickinson [3] and projected pumping through the 21st century based on population projections developed by the Arizona Department of Commerce [54] and TischlerBise [55], and calculating the total drawdown (Figure 3) and change in baseflow over the 1902–2100 period (Figure 11). These simulations predicted widespread increases in aquifer storage depletion across the western side of the basin during the 21st century (Figure 3b,c). They also quantified projected declines in baseflow in the basin over the next 100 years due to pumping and evapotranspiration, as well projected increases in baseflow until about 2050 resulting from the recharge facility at the City of Sierra Vista's Environmental Operations Park (EOP) (Figure 11b).

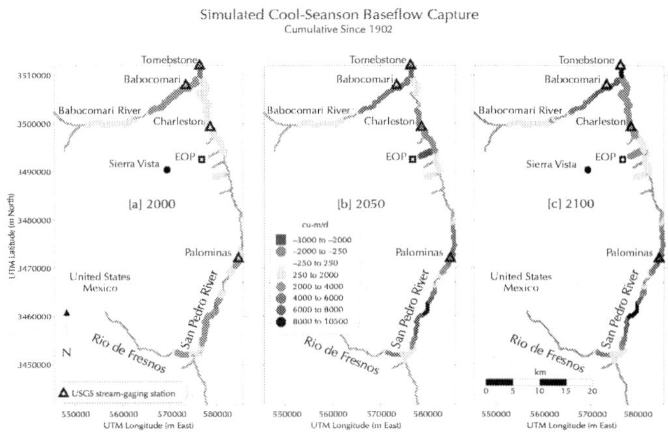

Figure 11. Simulated baseflow capture from 1902 to: (**a**) 2000; (**b**) 2050; and (**c**) 2100. Capture is cumulative [33] and measured in cu-m/d. EOP indicates location of Sierra Vista's Environmental Operations Park wastewater recharge facility.

Water budget calculations and groundwater simulations suggest that cumulative total groundwater depletion is presently on the order of 800 MCM in the subwatershed with annual net storage loss over the last decade on the order of 5 to 7 MCM [56,57]. Policy makers in the subwatershed have considered the feasibility of importing water at a maximum rate of roughly 37 MCM/yr [38]. However, even that most optimistic rate of importation would require nearly 25 years to bring the subwatershed's cumulative water budget back into balance. In the meantime, ongoing pumping-induced baseflow capture would continue to depress the groundwater gradient between the pumping centers and the river (Figure 4) further reducing baseflows.

Near-Stream Simulated Recharge Site Selection

Although artificial recharge of urban-enhanced runoff through detention basins has long been considered a viable option for mitigating impacts of groundwater pumping, the basins have previously been designed to target the active cone of depression in the Sierra Vista/Fort Huachuca area [34] rather than the river. The concept of simulating targeted near-stream recharge arose from the process of quantifying the pumping-induced capture of baseflow (*i.e.*, a reduction of groundwater discharge from the regional aquifer to the river) over time using the regional groundwater flow model [33]. With simulated current baseflow capture for the entire basin and projections of aquifer storage and baseflow capture trends over the next century, near-stream recharge was identified as a potential mechanism for addressing projected baseflow "deficits" (declines from 2003 (end of transient model calibration period) baseline values) in targeted areas of the basin.

Taken together, the suite of hydrologic tools pointed to sections of the San Pedro River where greater protection was warranted and where meaningful impacts—such as converting an intermittent stream reach into perennial reach—could be made. Simulations with the groundwater model [58] suggested that developing a distributed, strategically located network of recharge projects near select river reaches might result in groundwater mounding that would effectively compensate for baseflow capture, thus, protecting baseflow and the riparian vegetation community from the anticipated effects of pumping for several decades or more.

Application of Recharge Models

A small core team of technical experts within the Upper San Pedro Partnership Technical Committee selected three trial sites for simulating hypothetical recharge near the river: (1) Palominas; (2) Garden Canyon; and (3) Babocomari (Figure 12). The site selection process was informed by both simple geography (upper, middle, and lower part of the subwatershed) and current and projected baseflow capture in the river system. Simulated recharge at each trial site consisted of surface recharge over four 250 m × 250 m model cells (0.25 km^2) for the period 2012–2111. For each of the three trial sites shown in Figure 12, the recharge simulation investigation involved increasing recharge, as needed, to prevent any decline in baseflow below baseline (2003) levels downstream of the site over the simulation period while also preventing simulated surface flooding at the trial site.

Starting with 0.62 MCM/yr, simulated recharge at each site was incrementally increased, and baseflow response tested, until baseflow downstream of each site remained at or above 2003 levels for the 100-year simulation period. Figure 13 shows simulated change in the cool season (October–March) baseflow from 2003 (end of the transient calibration period) conditions in the years 2030, 2050, 2070, and 2111. As the decreasing cool (blue and green) colors in the northern half of the subwatershed over time indicate, baseflows are predicted to fall below 2003 levels in all of the mainstem San Pedro River and on the Babocomari by 2111. Recharge at the Sierra Vista EOP successfully maintains simulated baseflows in the mainstem above 2003 levels until at least 2070, but the impacts of pumping (deepening and widening cone of depression) overwhelm the recharge benefits by about the turn of the century.

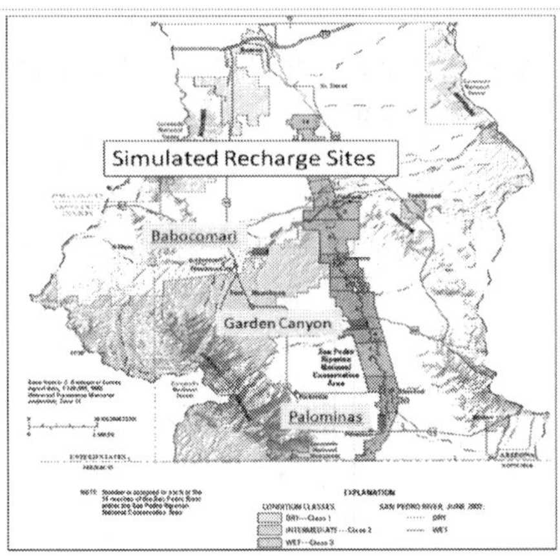

Figure 12. Simulated trial recharge sites (Babocomari, Garden Canyon, and Palominas) in the Sierra Vista subbasin. Riparian condition class reaches de-

lineated within the SPRNCA [10]. Adapted from Figure 42 in Stromberg and others [10].

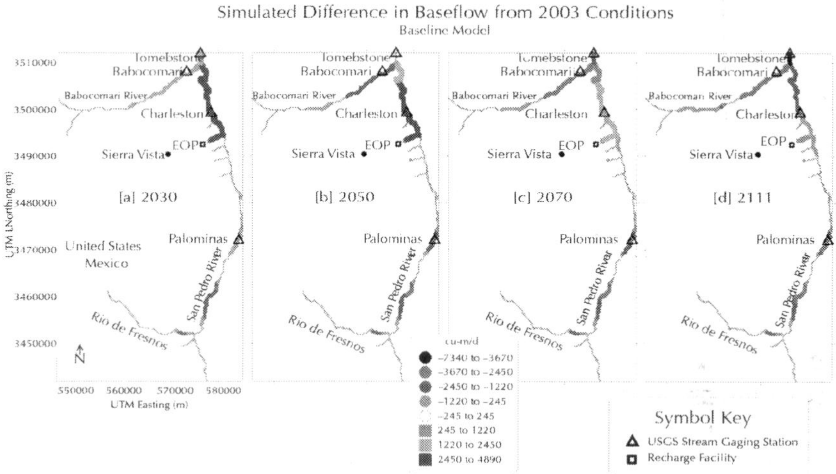

Figure 13. Simulated difference in baseflow (cu-m/d) in basin streams from 2003 conditions in: (a) 2030; (b) 2050; (c) 2070; and (d) 2111.

Simulation Results

Hydraulic conductivity, antecedent depth to groundwater, and projected aquifer storage depletion over time controlled the simulated baseflow response to recharge at each of the three trial sites. Since simulated recharge applied at each of the three trial sites was tailored to meet the anticipated baseflow deficit downstream, each site demanded a unique recharge distribution and exhibited a unique response to the simulated recharge. Figure 14 illustrates the recharge rates determined by trial and error as necessary to sustain simulated baseflows at or above 2003 levels in the groundwater model [3] through 2111. While each of the three test sites exhibited a unique response to the simulated recharge rates shown in Figure 14, the Babocomari site exhibited a much higher demand for recharge and a much more pronounced response in the underlying groundwater than the other two sites [58]. For the purpose of illustration, only the Babocomari test site recharge results will be discussed in detail here.

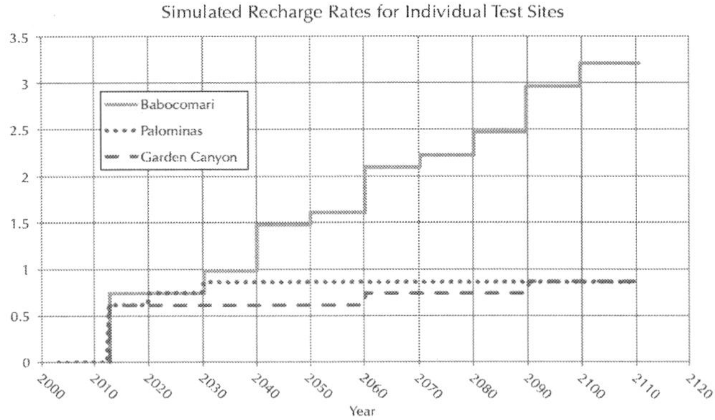

Figure 14. Simulated recharge rates for the Babocomari, Palominas, and Garden Canyon test sites.

Figure 15 illustrates the groundwater response to recharge at the Babocomari site. The black line represents depth to groundwater (DTW) under the stream adjacent to the recharge site in the baseline model with no simulated recharge at the Babocomari site. The orange line shows DTW in response to a constant rate of 0.62 MCM/yr. recharge, and the blue line shows DTW in the same location under the stream for the varying-recharge scenario illustrated by the green "Variable Recharge Rate" line. In the absence of any intervention (black DTW curve), simulated heads at this site are projected to drop by more than 10 m over the 21st century in response to pumping. However, these simulations suggest that incrementally increasing recharge from an initial rate of 0.74 MCM/yr in 2012 to 3.2 MCM/yr by 2100 would successfully maintain groundwater levels under the river at or slightly above the 2003 level.

The oscillation in the blue DTW curve reflects the fact that the simulated variable recharge increased groundwater levels under the river to a depth between the top of the evapotranspiration (ET) surface (1.5 m below top of aquifer) and the ET extinction depth at 6 m below the top of the aquifer. Thus, groundwater is more accessible to riparian vegetation in the varying-rate recharge scenario than in the baseline case, but baseflows still remain at or above 2003 levels in the area of the stream downstream where baseflow declines are projected under baseline conditions.

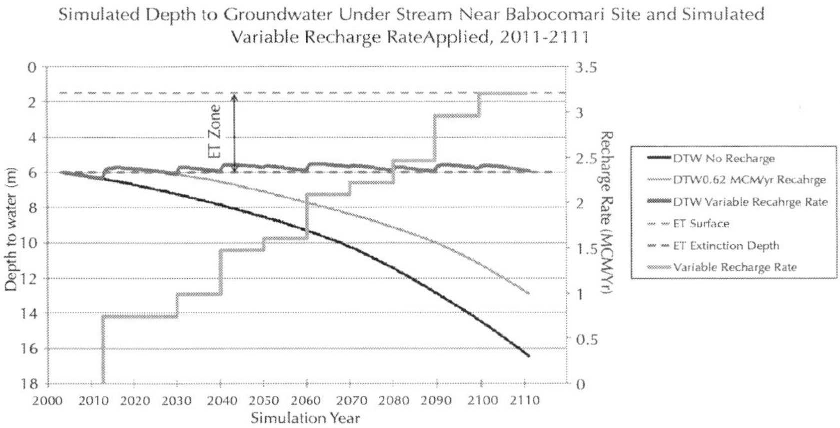

Figure 15. Simulated variable recharge rate at the Babocomari site and depth to water (DTW) under stream adjacent to the Babocomari recharge site from 2012 to 2111 for three scenarios: (a) no recharge; (b) 0.62 MCM constant-rate recharge; and (c) variable-rate recharge scenarios [58]. Evapotranspiration (ET) zone occurs between 1.5 and 6 m [3].

Optimization of Recharge Rates

Section 2.3.5 presented the results of simulated varying-rate recharge at each of three trial sites individually along the San Pedro and Babocomari Rivers. One additional simulation combining the three trial recharge sites was run in order to answer the question of whether hydrologic efficiencies might be gained with simultaneous recharge at all three sites [58]. Because the reach of the San Pedro affected by simulated recharge at the Garden Canyon site is downstream of both the Babocomari and Palominas sites, some reduction in the required recharge at the Garden Canyon site was achieved by combining all three trial recharge sites in a single simulation. Figure 16 illustrates the reduced recharge required at the Garden Canyon site to maintain simulated baseflows downstream of that site at 2003 levels when recharge is simulated at the Palominas and Babocomari trial sites concurrently.

Total recharge required to maintain simulated baseflows at 2003 levels downstream of each of the three test sites is shown in Figure

17. The blue curve shows the total recharge required when the three sites are simulated independently of each other, and the green curve shows the total recharge requirement when all three test sites are simulated concurrently. The difference between the curves illustrates the efficiency gained by combining the three recharge test sites. The maximum total recharge rate in the independent simulations is 4.93 MCM/yr (3.35 MCM/yr average over the 2012–2111 period), but that value drops to 4.63 MCM/yr (2.91 MCM/yr average) for the concurrent recharge simulations. The average recharge saved by operating all three sites concurrently is 0.45 MCM/yr. Figure 18 shows the final simulated change in baseflow from 2003 conditions when recharge at the three test sites is optimized for concurrent simulation of the three sites.

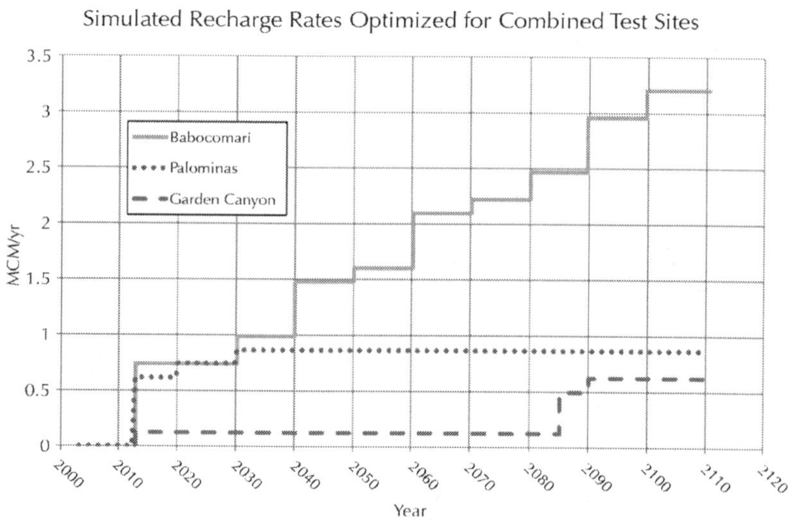

Figure 16. Simulated variable recharge rates for the Babocomari, Palominas and Garden Canyon test sites. Simulating recharge at all three sites concurrently allowed a reduction in Garden Canyon recharge relative to the rates for the independent recharge simulations.

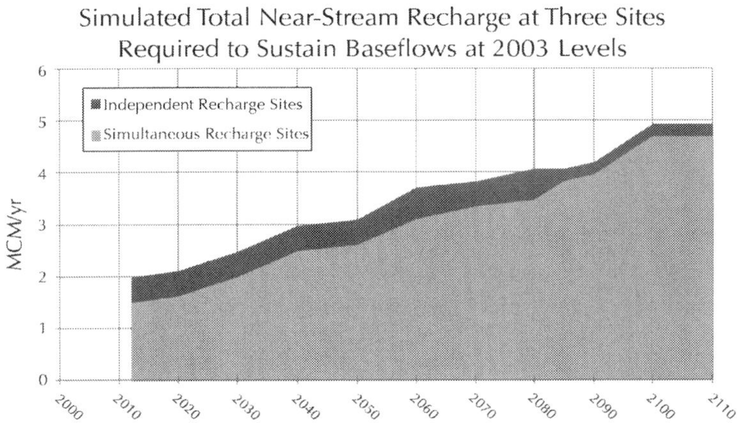

Figure 17. Total simulated recharge at the Babocomari, Palominas, and Garden Canyon test sites required to maintain baseflows downstream of each site at or above 2003 levels when recharge is simulated at each site independently and when all three sites are simulated concurrently.

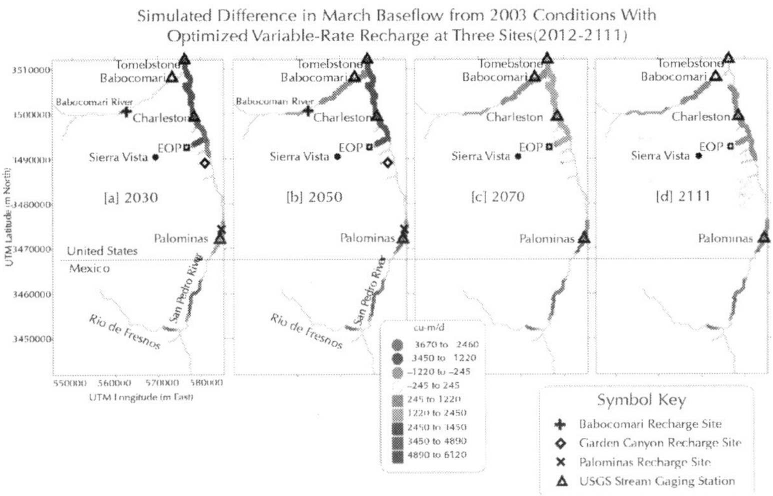

Figure 18. Simulated difference in March baseflow (cu-m/d) from 2003 conditions with optimized variable-rate recharge at three trial recharge sites in: (a) 2030; (b) 2050; (c) 2070; and (d) 2111.

Advantages/Disadvantages and Evolution of the Tool

Any perceived bias in the construction of the model, or simple disagreements among stakeholders with the technical modeling approach, can be problematic in terms of how modeling results will be used, if at all. Engagement of stakeholders throughout the model development process is essential for it to be embraced as a useful tool for decision making among varied interests. In this case, the groundwater model [3] was developed in response to wide dissatisfaction with some precursor models of the subwatershed. Richter and others [1] describe the process by which stakeholders were involved in the development of the Pool and Dickinson model [3] and the evolution of trust in the model among technical water resources experts and politicians in the basin.

Several years of experience with the Pool and Dickinson model [3] in developing and communicating results of baseline projections helped pave the way for using it as a tool to evaluate the prospects of near-stream recharge. The model made experimenting with various recharge rates and tracking the resultant changes in baseflow a fairly quick and inexpensive undertaking. While we feel that these simulation efforts successfully conveyed the general concept and potential merits of near-stream recharge to the public and decision makers, significant criticism arose from our choice of an arbitrary initial recharge rate of 0.62 MCM/yr for the hypothetical recharge sites. We viewed the initial near-stream recharge simulations as an exploratory mission to determine how much water would be required to produce the desired effect on baseflows in the San Pedro River, irrespective of the potential feasibility of attaining that quantity of water or distributing it in the locations of interest. As our initial simulation results were presented to stakeholders, many of them questioned the value in simulating recharge with water that is not available and for which no plans to develop were pending. We saw the near-stream modeling process as a "proof-of-concept" effort, but others quickly made the leap to the real difficulties in securing water for recharge. Significant time and energy were expended in efforts to bridge this conceptual gap, and in hindsight, more early effort to clarify the purpose and strategy behind the simulations would have been helpful.

Application of Recharge Simulations to Upper San Pedro Parcels

The primary value in the groundwater modeling efforts undertaken for the three "test sites" described above was in proving the potential benefit of multiple recharge sites near the river. In recent years, the U.S. Army Compatible Use Buffer program, Cochise County, and The Nature Conservancy collaboratively acquired and set aside for conservation purposes four parcels on the west side of the San Pedro River totaling 2226 hectares within the subwatershed (Figure 5). Collectively, these properties make up the physical sites currently under consideration for development of a network of near-stream recharge projects. The groundwater modeling process detailed above for the three hypothetical "test sites" was employed to some extent in the preliminary planning process for parcels 1 and 3 in Figure 5. As projects move from the conceptual phase to the physical site investigation stage, use of the groundwater model is being adapted to suit the project planning needs for the individual sites. While development of a recharge project on parcel 1 was constrained by several factors (parcel size, flood-control objectives, location of high-permeability soils off site), the current planning process for parcel 3 is relatively unconstrained. Modifications to model structure to reflect observed field conditions and refinement of the model grid to allow for more detailed simulation of potential recharge are two of the anticipated outcomes of the next phase of investigation at parcel 3.

RESULTS AND DISCUSSION

The suite of analytical tools discussed here is being used to inform key decisions necessary to balance groundwater use and maintain San Pedro River flows and associated riparian area ecological health. An extensive collection of hydrological studies and a robust, long-term monitoring program in the San Pedro basin have provided policy makers and stakeholders with important information about the complex relationships between groundwater condition, streamflow, and the ecological integrity of the riparian system within and near the SPRNCA. As a result, most of the stakeholders in the subwatershed understand that much of the San Pedro's riparian vegetation uses groundwater from the stream alluvium, and that this alluvial aquifer stores water from flood flows, receives groundwater from the regional aquifer, and contributes baseflow to the river during low-flow periods.

The groundwater model used in this study does not incorporate the complex interactions between flood-driven recharge of the shallow alluvium which influences baseflow to varying degrees from year to year, and pumping-induced depletions of the regional aquifer, which take decades to centuries to alter baseflow. While the model's authors made every effort to exclude storm-flow influences from the baseflow measurements they used in model calibration [3], it is virtually impossible to ensure that any given measured baseflow value represents only regional aquifer groundwater. The implication of this limitation is that the model may slightly overestimate the regional aquifer's contribution to baseflow, which would manifest as overestimated aquifer transmissivity and/or streambed conductance. That outcome, in turn, means that the model would also overestimate the simulated impact of pumping on baseflow until capture reaches its maximum (equilibrium) value. For the purpose of this study, which is to develop strategies for mitigating pumping-related impacts on baseflow, that response would be conservative and acceptable.

The Pool and Dickinson [3] model represents the culmination of years of study and is the best tool currently available for the study area. As discussed by Richter and others [1], using a groundwater model that is accepted by the vast majority of decision makers to perform predictive groundwater modeling has been essential for beginning to make management and project decisions from a common starting point. The modeling has demonstrated that within the next 100 years, two regional cones of depression will enlarge and likely change the nature of the hydrologic connection between the San Pedro River and the regional aquifer, reducing baseflow and impacting the dependent riparian system and thus wildlife populations.

Awareness and acceptance of an impending problem, however, is only the first step in finding a solution. The additional tools of the wet/dry maps and groundwater capture/recharge maps helped to focus management attention on finding both the most vulnerable areas of the system and the most beneficial locations for mitigation efforts. Analysis of wet/dry maps showing surface water presence during the driest time of year and areas with high year-to-year variation in wetted length may be the first physical evidence of changes in local groundwater conditions at the river. While wetted length is not solely controlled by groundwater conditions (it may also be affected by climate), different trends in various reaches of the river may help identify areas at higher risk of future ecological changes. Aligning these low-flow river reaches with the groundwater capture maps provided a rough indication of the rate that recharge in those areas of the aquifer might respond, expressing itself as baseflow in the San Pedro River. The capture maps also suggested the suitability of various locations for recharge that may communicate with the alluvial aquifer and/or the San Pedro River. On-site investigation of actual hydrogeologic conditions and suitability for recharge at a particular location is the first step toward refining a preliminary conceptual model derived from the tools described in this paper.

In the San Pedro basin, the use of these complementary approaches informed the purchase of the most hydrologically sensitive lands near or adjacent to SPRNCA (Figure 5) in order to both defer residential and/or agricultural development and provide the opportunity for near-stream recharge project development. The concept of a strategically located network of recharge projects near these river reaches evolved, in part, from the success of more than a decade of managed aquifer recharge at the City of Sierra Vista Environmental Operations Park in supporting both baseflow and replenishing the deep regional aquifer. We anticipate that recharging urban-enhanced runoff, storm water, and treated effluent near at-risk reaches shown would create groundwater mounds to sustain surface water flow and supplement alluvial groundwater levels during low-flow periods, effectively compensating for the otherwise deleterious impacts of encroaching cones of depression in the regional aquifer on baseflows for the next several decades or more.

CONCLUSIONS

The application of the study methods presented in this paper and the development of system-specific techniques appear to have great promise for protecting dry-land riparian systems from the impacts of groundwater extraction, surface water diversions, and the extremes of climate change for up to several decades. The ecological, cultural, and economic significance of the San Pedro River has made it one of the most well-studied and understood river systems in the world. The tremendous volume of data and hydrologic tools already developed for this specific system coupled with many years of collaborative partnerships that have matured with the science make the San Pedro basin a very unique policy environment. There is likely no other river system with an identical set of social, political, and ecological circumstances, but the hydrologic analysis tools described in this paper can be used anywhere. The strength of collaborative partnerships and knowledge of which tools they jointly support is key to building a common understanding of the history, goals, and resources at hand to make real progress.

ACKNOWLEDGMENTS

The authors would like to acknowledge the Walton Family Foundation for its support in funding much of the science and policy development in the Upper San Pedro Basin in recent years. They would also like to express gratitude to the reviewers whose collective input substantially improved this paper.

REFERENCES

1. Richter, H.E.; Gungle, B.; Turner, D.S.; Bushman, B.; Lacher, L.J. Development of a Shared Vision Toward Sustainability of Groundwater to Protect and Sustain Baseflows of the Upper San Pedro River, Arizona. *Water* 2014, 6, 2519–2538.
2. United States Congress. Arizona-Idaho Conservation Act of 1988. *Public Law 100-696*, Title I. 1988.
3. Pool, D.R.; Dickinson, J.E. *Ground-Water Flow Model of the Sierra Vista Subwatershed and Sonoran Portions of the Upper San Pedro Basin, Southeastern Arizona, United States, and Northern Sonora, Mexico*; U.S. Geological Survey: Reston, VA, USA, 2007; p. 47.
4. Cooperative Climatological Data Summaries for Arizona. Data for Coronado National Monument, Sierra Vista, and Tombstone

stations. Available online: http://www.wrcc.dri.edu/summary/Climsmaz.html (accessed on 13 June 2014).

5. Thomas, B.E.; Pool, D.R. *Trends in Streamflow of the San Pedro River, Southeastern Arizona, and Regional Trends in Precipitation and Streamflow in Southeastern Arizona and Southwestern New Mexico*; U.S. Geological Survey Professional Paper 1712. U.S. Geological Survey: Reston, VA, USA, 2007; p. 79.
6. 2013–2050 Sub-county Population Projections. Available online: http://www.workforce.az.gov/population-projections.aspx (accessed on 13 June 2014).
7. July 1, 2013 Population estimates for Arizona's counties, incorporated places and unincorporated balance of counties. Available online: http://www.workforce.az.gov/pubs/demography/July%201_2013%20PopulationEstimates_Final.pdf (accessed on 13 June 2014).
8. Carreira, R. Economic Impact of Fort Huachuca. Undated white paper. Center for Economic Research at Cochise College. Available online: http://www.cochise.edu/cfiles/files/CER/CER%20Studies/Economic_Impact_of_Fort_Huachuca.pdf (accessed on 13 June 2014).
9. Skagen, S.K.; Melcher, C.P.; Howe, W.H.; Knopf, F.L. Comparative use of riparian corridors and oases by migrating birds in southeast Arizona. *Conserv. Biol.* 1998, *12*, 896–909.
10. Stromberg, J.C.; Lite, S.J.; Dixon, M.; Rychener, T.; Makings, E. Relations between streamflow regime and riparian vegetation composition, structure, and diversity within the San Pedro Riparian National Conservation Area, Arizona. In *Hydrologic Requirements of and Consumptive Ground-Water Use by Riparian Vegetation along the San Pedro River, Arizona*; Leenhouts, J.M., Stromberg, J.C., Scott, R.L., Eds.; U.S. Geological Survey Scientific Investigations Report 2005-5163. U.S. Geological Survey: Reston, VA, USA, 2006; pp. 57–106.
11. Stromberg, J.C.; Bagsted, K.J.; Leenhouts, J.M.; Lite, S.J.; Making, E. Effects of stream flow intermittency on riparian vegetation of a semiarid region river (San Pedro River, Arizona). *River Res. Appl.* 2005, *21*, 925–938.
12. Haynes, C.V., Jr. Curry Draw, Cochise County, Arizona: A late Quaternary stratigraphic record of Pleistocene extinction and

paleo-Indian activities. In *Geologic Diversity of Arizona and Its Margins: Excursions to Choice Areas*; Arizona Bureau of Geology and Mineral Technology Special Paper 5. Davis, G.H., VandenHolder, E.M., Eds.; U.S. Department of the Interior: Phoenix, AZ, USA, 1987; pp. 23–28.

13. U.S. Bureau of Land Management. Tucson Field Office Paleontology Sites. Available online: http://www.blm.gov/az/st/en/prog/cultural.html (accessed on 15 June 2014).
14. Hereford, R. *Entrenchment and Widening of the Upper San Pedro River, Arizona*; Special Paper 282. Geological Society of America: Boulder, CO, USA, 1993; p. 46.
15. Arizona Department of Water Resources. Surface Water Adjudications. Available online: http://www.azwater.gov/AzDWR/SurfaceWater/Adjudications/ (accessed on 15 June 2014).
16. U.S. Department of Agriculture—Agricultural Research Service Experimental Watersheds. Available online: http://www.ars.usda.gov/Research/docs.htm?docid=10978&page=2 (accessed on 15 June 2014).
17. Brown, S.G.; Davidson, E.S.; Kister, L.R.; Thomsen, B.W. *Water Resources of Fort Huachuca Military Reservation, Southeastern Arizona*; U.S. Geological Survey Water-Supply Paper 1819–D. U.S. Geological Survey: Reston, VA, USA, 1966; pp. D1–D57.
18. Brown, S.G.; Aldridge, B.N. *Streamflow Gains and Losses and Ground-Water Recharge in the San Pedro River Basin, Arizona*; USGS Administrative Report. U.S. Geological Survey: Reston, VA, USA, 1973; p. 45.
19. Freethey, G.W. *Hydrologic Analysis of the Upper San Pedro Basin from the Mexico–United States International Boundary to Fairbank, Arizona*; U.S. Geological Survey Open-File Report 82–752. U.S. Geological Survey: Reston, VA, USA, 1982; p. 52.
20. Vionnet, L.B.; Maddock, T. *Modeling of Ground-Water Flow and Surface Water/Groundwater Interactions in the San Pedro River Basin—Part I—Cananea, Mexico to Fairbank, Arizona*; HWR Report No. 92-010. University of Arizona: Tucson, AZ, USA, 1992; p. 96.

21. Corell, S.W.; Putman, F.; Lovvik, D.; Corkhill, F. *A Groundwater Flow Model of the Upper San Pedro Basin, Southeastern Arizona*; Phoenix, Arizona Department of Water Resources Modeling Report No. 10. Arizona Department of Water Resources: Phoenix, AZ, USA, 1996; p. 85.
22. Goode, T.; Maddock, T., III. *Simulation of Groundwater Conditions in the Upper San Pedro Basin for the Evaluation of Alternative Futures*; HWR report No. 00-30. Department of Hydrology and Water Resources, University of Arizona: Tucson, AZ, USA, 2000.
23. Arizona Department of Water Resources. *Hydrographic Survey Report (HSR) for the San Pedro River Watershed*; Arizona Department of Water Resources: Phoenix, AZ, USA, 1990; p. 548.
24. Arizona Department of Water Resources. *Upper San Pedro Basin Active Management Area Report, March 2005*; Arizona Department of Water Resources: Phoenix, AZ, USA, 2005.
25. Pool, D.R.; Coes, A.L. *Hydrogeologic Investigations of the Sierra Vista Sub-Watershed of the Upper San Pedro Basin, Cochise County, Southeast Arizona*; USGS Water Resources Investigations Report 99-4197. U.S. Geological Survey: Tucson, AZ, USA, 1999; p. 41.
26. Cooperative water use management related to Fort Huachuca, Arizona, and Sierra Vista subwatershed. *U.S. Public Law 108-136*, Section 321. 2003.
27. Upper San Pedro Partnership. *Water Management of the Regional Aquifer in the Sierra Vista Subwatershed, Arizona—2011 Report to Congress*; U.S. Department of Interior: Washington, DC, USA, 2013; p. 16.
28. Coes, A.L.; Pool, D.R. *Ephemeral-Channel and Basin-Floor Infiltration in the Sierra Vista Subwatershed, Arizona*; U.S. Geological Survey Open-File Report 05–1023. U.S. Geological Survey: Reston, VA, USA, 2005; p. 67.
29. Gungle, B. *Timing and Duration of Flow in Ephemeral Streams of the Sierra Vista Subwatershed of the Upper San Pedro Basin, Cochise County, Southeastern Arizona*; U.S. Geological Survey Scientific Investigations Report 2005–5190. U.S. Geological Survey: Reston, VA, USA, 2006; p. 47.

30. *Hydrologic Requirements of and Consumptive Ground-Water Use by Riparian Vegetation Along the San Pedro River, Arizona*; Leenhouts, J.M., Stromberg, J.C., Scott, R.L., Eds.; U.S. Geological Survey Scientific Investigations Report 2005-5163. U.S. Geological Survey: Reston, VA, USA, 2006; p. 154.
31. Leake, S.A.; Pool, D.R.; Leenhouts, J.M. *Simulated Effects of Ground-Water Withdrawals and Artificial Recharge on Discharge to Streams, Springs, and Riparian Vegetation in the Sierra Vista Subwatershed of the Upper San Pedro Basin, Southeastern Arizona*; U.S. Geological Survey Scientific Investigations Report 2008-5207. U.S. Geological Survey: Reston, VA, USA, 2008; p. 14.
32. Kennedy, J.R.; Gungle, B. *Quantity and Sources of Baseflow in the San Pedro River near Tombstone, Arizona*; U.S. Geological Survey Scientific Investigations Report 2010–5200. U.S. Geological Survey: Reston, VA, USA, 2010; p. 43.
33. Lacher, L.J. *Simulated Groundwater and Surface Water Conditions in the Upper San Pedro Basin, 1902–2105*; Lacher hydrological Consulting: Tucson, AZ, USA, 2011.
34. Stantec. *Cochise County Flood Control/Urban Runoff Recharge Plan*; Stantec Consulting, Inc.: Tucson, AZ, USA, 2006.
35. GeoSystems Analysis, Inc. *Hydrologic and Incidental Groundwater Recharge Study*; Stantec Consulting, Inc.: Tucson, AZ, USA, 2006; p. 70.
36. U.S. Department of Agriculture—Agricultural Research Station. Automated Geospatial Watershed Assessment Tool, developed in cooperation with the U.S. Environmental Protection Agency. Available online: http://www.tucson.ars.ag.gov/agwa/index.php?option=com_content&view=article&id=5&Itemid=26 (accessed on 14 June 2014).
37. GeoSystems Analysis, Inc. *Storm Water Recharge Feasibility Analysis, Appendix B: AGWA/KINEROS Simulations for Coyote Wash Watershed*; Report to Cochise County for Project SP-0011. GeoSystems Analysis, Inc.: Tucson, AZ, USA, 2004.
38. United States Bureau of Reclamation-Lower Colorado Region. *Appraisal Report: Augmentation Alternatives for the*

Sierra Vista Subwatershed, Arizona; U.S. Department of Interior: Washington, DC, USA, 2007; p. 77.

39. United States Bureau of Reclamation. Grant No. 00-FC-32-0030 Between Bureau of Reclamation Phoenix Area Office and The City of Sierra Vista Pertaining to Construction of the Sierra Vista Water Reclamation Facility Effluent Recharge Demonstration Project. February 2000.
40. Brown and Caldwell. *Sierra Vista Water Reclamation Facility Groundwater Modeling Study*; Prepared for City of Sierra Vista. Brown and Caldwell: Sierra Vista, AZ, USA, 2009.
41. Turner, D.; Richter, H. Wet/Dry Mapping: Using Citizen Scientists to Monitor the Extent of Perennial Surface Flow in Dryland Regions. *Environ. Manag.* 2011, *47*, 497–505.
42. Theis, C.V. The source of water derived from wells: Essential factors controlling the response of an aquifer to development. *Civ. Eng.* 1940, *10*, 277–260.
43. Lohman, S.W.; Bennett, R.R.; Brown, R.H.; Cooper, H.H., Jr.; Drescher, W.J.; Ferris, J.G.; Johnson, A.I.; McGuiness, C.L.; Piper, A.M.; Rorabaugh, M.I.; *et al. Definitions of Selected Ground-Water Terms—Revisions and Conceptual Refinements*; U.S. Geological Survey Water-Supply Paper 1988. U.S. Geological Survey: Reston, VA, USA, 1972; p. 21.
44. Bredehoeft, J.D.; Papadopulos, S.S.; Cooper, H.H. Groundwater: The water budget myth. In *Scientific Basis of Water-Resource Management, Studies in Geophysics*; National Academy Press: Washington, DC, USA, 1981; pp. 51–57.
45. Brown, R.E. *The Cone of Depression and the Area of Diversion around a Discharging Well in an Infinite Strip Aquifer Subject to Uniform Recharge*; U.S. Geological Survey Water-Supply Paper 154SC. U.S. Geological Survey: Reston, VA, USA, 1963; pp. C69–C85.
46. Winter, T.C.; Harvey, J.W.; Franke, O.L.; Alley, W.M. *Ground Water and Surface Water: A Single Resource*; U.S. Geological Survey Circular 1139. U.S. Geological Survey: Reston, VA, USA, 1998; p. 15.

47. Heath, R.C. *Basic Groundwater Hydrology*; U.S. Geological Survey Water Supply Paper 2220. U.S. Geological Survey: Reston, VA, USA, 1986; p. 86.
48. Lohman, S.W. *Ground-Water Hydraulics*; U.S. Geological Survey Professional Paper 708. U.S. Geological Survey: Reston, VA, USA, 1972; p. 70.
49. Leake, S.A.; Reeves, H.W.; Dickinson, J.E. A new capture fraction method to map how pumpage affects surface water flow. *Groundwater* 2010, *48*, 690–700.
50. Barlow, P.M.; Leake, S.A. *Streamflow Depletion by Wells—Understanding and Managing the Effects of Groundwater Pumping on Streamflow*; U.S. Geological Survey Circular: 1376. U.S. Geological Survey: Reston, VA, USA, 2012; p. 95.
51. Leake, S.A.; Greer, W.; Watt, D.; Weghorst, P. *Use of Superposition Models to Simulate Possible Depletion of Colorado River Water by Ground-Water Withdrawal*; U.S. Geological Survey Scientific Investigations Report 2008-5189. U.S. Geological Survey: Reston, VA, USA, 2008; p. 25.
52. Harbaugh, A.W.; Banta, E.R.; Hill, M.C.; McDonald, M.G. *MODFLOW-2000, the U.S. Geological Survey Modular Ground-Water Model—User Guide to Modularization Concepts and the Ground-Water Flow Process*; U.S. Geological Survey Open-File Report 00-92. U.S. Geological Survey: Reston, VA, USA, 2000; p. 121.
53. Prudic, D.E. *Documentation of a Computer Program to Simulate Stream-Aquifer Relations Using a Modular, Finite-Difference, Ground-Water Flow Model*; U.S. Geological Survey Open-File Report 88-729. U.S. Geological Survey: Reston, VA, USA, 1989; p. 113.
54. Arizona Department of Commerce (AzDC) population projections for 2006, 2009, and 2010. Available online: http://www.azcommerce.com/EconInfo/Demographics/Population+Projections.htm (accessed on 30 October 2010).
55. TischlerBise. *Infrastructure Improvements Plan and Development Fee Study*; consulting report prepared for the City of Sierra Vista. TischlerBise: Bethesda, MD, USA, 2009.

56. *Upper San Pedro Partnership—Water Management of the Regional Aquifer in the Sierra Vista Subwatershed, Arizona—2004 Report to Congress*; U.S. Department of Interior: Washington, DC, USA, 2005; p. 36.
57. *Upper San Pedro Partnership—Water Management of the Regional Aquifer in the Sierra Vista Subwatershed, Arizona—2010 Report to Congress*; U.S. Department of Interior: Washington, DC, USA, 2011; p. 20.
58. Lacher, L.J. *Simulated Near-Stream Recharge at Three Sites in the Sierra Vista subbasin, Arizona*; Lacher hydrological Consulting: Tucson, AZ, USA, 2012.

Chapter 7

A Haitian 'Ecodistrict:' Conceptual Design for Integrated, Basic Infrastructure for the Commune of Léogâne

Hillary Brown[1], and Miriam N Ward[2]

[1]The Spitzer School of Architecture, The City College of New York, 141 Convent Avenue 3 M09, New York, NY 10031, USA

[2]Sustainability in the Urban Environment Program, The City College of New York, 160 Convent Avenue, New York, NY 10031, USA

ABSTRACT

Background

The island nation of Haiti poses an ongoing challenge to the international development community. Political instability and lack of basic public services, combined with weak technical and financial capacity, collectively undermine both philanthropy and international investment. Haiti's tragic vulnerability to natural disasters (earthquakes and hurricanes) and the public health crises that ensue have only compounded its level of destitution. These complex, interconnected circumstances doom single-threaded solutions.

Methods

The problems call instead for integrative responses, answers that cut across sectors and address multiple problems simultaneously. This proposal for critical services development in the Haitian commune of Léogâne seeks to reduce much of its poverty, malnourishment, pollution, health problems, flooding, deforestation, topsoil erosion and loss of biodiversity.

Results

The territory's long degraded ecosystem services, once regenerated through reforestation and riverbank stabilization, will boost agricultural production. More significantly, they will help provision targeted new energy, water and waste management systems designed together with synergistic exchanges in mind.

Conclusions

Across these sectors shared resource and energy flows will improve efficiencies, lower costs and reduce environmental impacts.

BACKGROUND

Set against decades of political volatility and economic stagnation, Haiti's marginal investments in critical infrastructural services have left its people struggling today with growing population pressures and increased needs in a context of dwindling resources. Haiti was already the poorest country in the Western hemisphere, with an unemployment rate as high as 70%, Adelman (2011) when its capital was devastated by a magnitude 7.0 earthquake on January 12th of 2010. The tragedy was compounded by non-existent building codes and shoddy construction: somewhere between 230,000 to 316,000 Haitians lost their lives beneath collapsing structures, with hundreds of thousands more injured (O'Connor 2012). Port-au-Prince, which houses major government and administration offices, was devastated. With so many of the nation's public servants buried in imploding buildings, basic services were disabled for the long months that followed. Globally, well-meaning nongovernmental organizations (NGOs) rushed to offer services and to rally a hobbled people. But Haiti, already a nation with the highest number of NGOs per capita, was overwhelmed by what became a disorganized aid effort. Miscommunications and the lack of coordination that ensued only added to the chaos (Klarreich & Polman 2012). The government worked with international organizations to clear rubble and restore necessary services, while trying to rebuild internal capacity with limited funds. In October 2010, the unthinkable happened again, when a cholera outbreak swept through regions lacking energy, sanitation and secure potable water. As of May, 2013, the epidemic had claimed more than 8,000 lives (Cholera in Haiti). Ambitious plans to "Build Back Better," (disaster recovery plans undertaken by the Clinton Foundation, along with many others) have unfortunately stagnated. At this writing, three years following the quake, the Haitian people find themselves with only spotty improvements, increased need, and declining opportunities (Sontag 2012).

Current Context

At the epicenter of the earthquake, the commune of Léogâne (population then, 130,000) sustained severe losses, with an estimated 93% of its buildings damaged (Eberhard et al. 2010) (Figure 1). While rubble has been substantially cleared, the few services that existed remain unrestored. Critical infrastructure needs to be comprehensively rebuilt. This poses an opportunity to consider progressive paradigms for utilities' integration and management.

Figure 1: Haiti overview.

Official waste collection in Léogâne is non-existent. Most refuse is piled and periodically set on fire, a form of 'local management (University of Notre Dame 2011a).' Plastics, along with other household debris, are thrown askance and washed downstream, clogging drainage channels and canals. The town lacks sanitary systems for human and animal waste, and waterways are overrun with organic material, creating serious health hazards (Archibold 2012).

The existing electrical grid (servicing only a small portion of the community) has not functioned since the earthquake. While the government has promoted a plan to connect Léogâne to the national grid, nothing has transpired to date (University of Notre Dame 2011b). In July 2012, Electricité d'Haïti (EDH) and the Ministry for Public Works partnered with Korea International Cooperation Agency (KOICA) to install a small microgrid serving 20,000 Léogâne residents, powered by diesel generators (Haiti - Energy 2012). The system is still not yet fully functional. Clinics, organizations and families with resources are left to run private generators with expensive, imported fuel. Few have been able to purchase solar panels. Throughout Haiti, lack of affordable energy limits employment, buying-power, access to education, opportunities, and investment capacity (Barjon 2011a). For a nation largely comprised of remote rural poor, a distributed energy plan realized with local natural resources could greatly improve the standard of living for much of Haiti's population (OECD 2006a).

The water distribution in Léogâne is similarly problematic. During the 2008 hurricane season, the underground water supply system serving the town center was damaged and, without funding for repair, it has not functioned since (Piasecki 2013). Much of the population still struggles to access potable water. Those with means utilize private wells, while those without utilize shallow hand-dug wells, walk to a public water station or buy water in costly plastic pouches or other containers which are subsequently discarded. While the local water authority, DINEPA, has plans for restoring supply pipelines to the center, service to peripheral and rural areas will be limited to water kiosks (Rebuilding Léogâne Planning Workshop 2011a). The drainage system had been in a similar state of disrepair, but has recently been repaired through international aid. Much of the system, however, lies below sea level. Its drainage cannot be discerned, nor are the parties involved sharing operational plans (Piasecki 2013). Without transparency and local involvement progress is limited.

Transportation services are minimal and crude. Léogâne's limited roadway system is in severely deteriorated condition (University of Notre Dame 2011c). Market activities choke the town's central streets, limiting vehicle volume and speed (Shepard 2010). Roadbeds, rutted with puddles, have become vectors for mosquito nesting and water-borne illnesses. (Léogâne is a known center for mosquitos carrying lymphatic filariasis, which causes Elephantiasis.) Rural roadways are unpaved and intermittently overwhelmed by deluges, increasing the extent of erosion. To manage flooding, open drainage ditches flank roadways, posing significant hazards to unwary drivers. The lack of safe roadways also limits deployment of heavy equipment.

Large areas across Haiti have experienced severe environmental degradation. Overall, only 2% of the country remains forested

(Library of Congress –Federal Research Division 2006). Both upland trees and coastal mangroves in this energy-impoverished landscape have been widely harvested for charcoal, the primary cooking fuel. Severe deforestation is also endemic in and around Léogâne, along the coastal areas and into its surrounding hills and mountains. No longer stabilized by root systems, upland soil erodes during tropical downpours. It overwhelms rivers and streams, fanning out downstream as it washes into the ocean. Routine flooding takes out roadbeds and creates impassable conditions across the alluvial plain in and around the town center.

Tropical floods have decimated the once confining banks of the Momance and Rouyonne Rivers, severely damaging not only riparian habitat, but also the deltas and estuaries. While natural causes exist for coastal habitat degradation (including the earthquake's seismic uplift that drained tidal marsh, exposed sub-tidal sea grass causing loss of coral reef and mangrove habitat), (Koehler & Mann 2011) anthropogenic impacts compound the damage through the massive influx of river sedimentation and the harvesting of mangroves for fuel (Rebuilding Léogâne Planning Workshop2011b). Sited within the Caribbean's well-trafficked "Hurricane Alley," Léogâne's developed land area, along with its once bio-rich estuaries and coastal coral reefs, has become especially vulnerable to rising sea levels. Earthquake damage, coupled with environmental degradation, has triggered negative feedback loops, limiting local economic recovery. Practical opportunities for combined solutions to these interconnected problems have driven the integrative design of this proposed ecodistrict. (Ecodistricts are an urban planning term for sustainable development in a larger region where the ecological footprint is considered in planning and resources are used judiciously (Metzger and Olsson 2013)).

New Paradigms for Utilities in Developing Nations

Sixty percent of global energy growth in the next few decades will take place in developing nations, where about two billion people are living without access to electricity (Holm & McIntosh 2008). Many emerging economies, while constrained by other substantial barriers to progress, enjoy a potential window of opportunity to bypass petroleum-dependent technologies for next generation systems. Through implementation of largely renewable energy-driven solutions upfront, these nations may accelerate the technology application in holistic ways. For example, smart-grid and distributed micro-grid networks, compared to centralized systems, reduce losses while better matching load with demand (Gregory 2011a). By investing in renewable energy sources, developing economies may draw roadmaps for a lower carbon, global energy transition. This opportune, but non-traditional pathway, however, places significant burden of risk on energy policy decision-makers. Leapfrogging fossil fuel technologies will also entail considerable support and international cooperation from both governmental and private investment communities (Holm 2005). Globally, carbon-based energy systems have enjoyed inter-governmental protection, subsidy policies, and economies of scale achieved through investment in conventional generation and transmission (Holm & McIntosh 2008).

GEMi and a Vision for Haiti

Project objectives and strategies for this ecodistrict design build upon, and are aligned with, the precepts of the Global Energy Model (GEM). The Global Energy Model Institute, a new nongovernmental initiative that bears its name (GEM Institute or GEMi), was formed to address many of these challenges (Global Energy Model Institute 2012). Its broad objective is to solve global energy poverty through the progressive application of clean, reliable power systems, based on proven technologies. GEM is the organizing framework for multi-sector investment and development with energy at its hub. Haiti is its initial pilot. A premise for its universal application is the nature of complex, cross-sectoral interactions around, and dependencies upon, energy. While GEM is technologically agnostic, it is biased towards solutions that serve the triple bottom line. The energy model itself is an integrated information management system that captures innovations and best practices and allows for successive refinement across different locales (Under development by Daniel C. Gregory of Positive Energies, LLC).

For Léogâne, it was assumed that many of GEMi's basic operating principles and implementation processes would be applied as givens. These include the reliance on stakeholder-driven development,

development of Haitian professional and administrative capacity, and the utilization, wherever possible, of a local work-force. GEMi's mission also entails enabling and locally sustaining an expanding power system infrastructure through its planned education and training programs (Gregory2010).

Haiti lacks a national grid beyond the one that currently serves Port-au-Prince and immediate environs. One of the visions for Haiti elaborated upon through GEMi involves the development of a national grid based significantly on existing hydropower that will be upgraded and enhanced (Brown2011). For locales off the planned primarily North–south grid, a decentralized generation system is envisioned, with eventual grid connectivity. (Lemons, 2012). With Haiti's island context, renewable energy generation options are ultimately more cost-effective and viable due to the logistical and economic complications of foreign fossil fuel delivery and supply.

GEMi's reliance on diversified and networked renewable resources reduces dependence on imported fuel, limits carbon emissions and provides system diversification and resiliency. Currently, Haiti's energy is largely sourced from foreign fossil fuel hydropower; however, the existing dams are undergoing upgrades to increase the grid's base-load capacity. Refurbishing the dams risks extending negative impacts to the remaining intact ecological services contributed by the river. Sedimentation from unmanaged upland erosion must be constantly removed to maintain capacity. Moreover, methane, a highly potent greenhouse gas, is also emitted by untended decaying biological matter. Other options pose technical challenges. Both wind power and solar energy production fluctuate. Neither alone is sufficiently reliable, and each would require some measure of storage to be able to release energy on-demand. New multipronged or hybrid systems such as blended renewables, coupled with storage and backed up by diesel-fueled generation are envisioned to address the breadth of the challenge.

METHODS

Holistic Design for Synergies across Sectors: Léogâne as Model for Rural Utilities

Plans for Léogâne's redevelopment include environmental remediation and optimization of material and resource flows across the energy, water, agricultural and transportation sectors. Investment in one sector, per the plan, creates the stability needed to initialize another; this is integral to the final outcome and can't be undertaken successfully unless the supporting sectors are in place. The new paradigms for the plan rely upon the cascading of energy, water or residual resources extracted from nature across these sectors. Through colocation and coupling of complementary components, the program will capitalize on designed-in exchanges among multiple system elements. For instance, floodwater control may also be used to generate energy, irrigate crops and improve water quality. Agricultural and municipal solid waste can provide energy and heat for industrial processes while returning beneficial soil amendments. The vision entails closing loops to create a self-sufficient, holistic system, one that captures synergies, creates economic efficiencies, and eliminates waste. Promoting synergies between economic gains and social infrastructure helps to amplify benefits, ensuring that real gains are made in all sectors for the poorest and most vulnerable people (OECD2006b). The design also promotes colocation of complementary entities such as industry, hospitals, clinics and schools, allowing these entities to take advantage of infrastructural outputs and/or contribute inputs. Overall, the plan reduces reliance upon most 'first world' carbon-intensive systems, substituting for the most part technologies that can, after training, be constructed and managed by local labor. In addition to new jobs, the scheme also entails alternative employment opportunities for agriculture, agroforestry, and aquaculture industries envisioned for the commune, along with other forms of local enterprise.

Léogâne is an ideal locus for reconstruction as its landscape mirrors that of much of Haiti, with its steep rural mountains, rivers and streams, and its populous alluvial plains. Successes in Léogâne could be used as templates for integrated development across rural

Haiti where conditions are similar. Interventions in Léogâne were therefore designed along just such a typical corridor running from the mountainous uplands in the South, through the more populous alluvial plain and down to the coast on the North. The area is bounded on either side by the Momance River and the Rouyonne River, both of which have large seasonal variations.

Local efforts to guide reconstruction along this mountain-to-coast transect will help to inform placement of the system's modular elements. Working components will be sited at locations with specific relationships to both natural resources and existing settlements (Figure 2). For example, the upland placement of wind turbines captures higher wind speeds. Photovoltaic arrays, interspersed with agriculture, occupy the middle range and power the pumping of water from a nearby receiving reservoir to a holding reservoir above for the pumped-hydro-storage generation system. Within this region too, biomass salvaged from agriculture will be used to supplement energy generation. All along this corridor, riparian resources will be stabilized for flood control, infrastructure protection, and for improved water quality for farming and aquaculture.

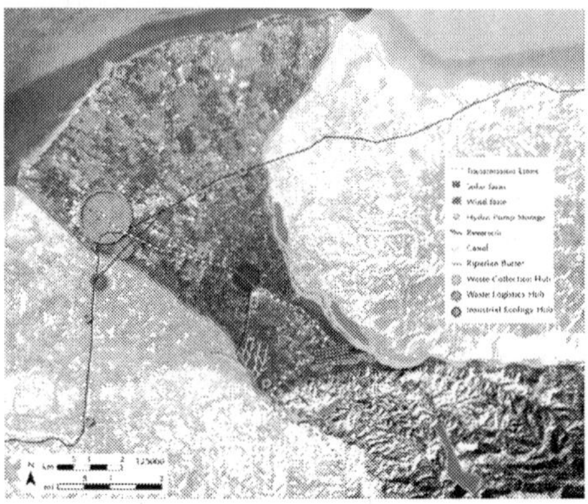

Figure 2: Integration of critical infrastructure overview.

A base load generated by a pumped-hydro storage system (PHS) will power the industrial, civic and commercial sector. Explained below, the PHS is a generation system networked for reliability and redundancy. The distribution of electrical services also relies on an integrated delivery mechanism: a set of service points or "nodes" (Figure 3). This approach standardizes and maximizes the reach of new services. In the town center, for example, nodes serve neighborhoods, public plazas, the markets and other civic functions. Nodal locations will be linked by a newly paved road network. These stations will also serve as outlets for clean drinking water and as drop-off locales for waste collection. Here, cellphones may be recharged, as well as auxiliary LED lamps. Organic and solid waste dropped off at these service points (incentivized through rebates) will be carted to a nearby waste processing (bio digestion) plant site for augmentation of the power-supply. Economies obtained from colocation and simultaneous construction of these services will make the best use of limited resources until more are at-hand to finally extend services to individual dwellings. The nodal service points create an integrated web of services linking new and existing infrastructure. A few are designed to incorporate auxiliary community services such as "electronic cafes." A number of "nodes as community-hubs" are also planned for the town square, on school campuses, clinics and at the hospital.

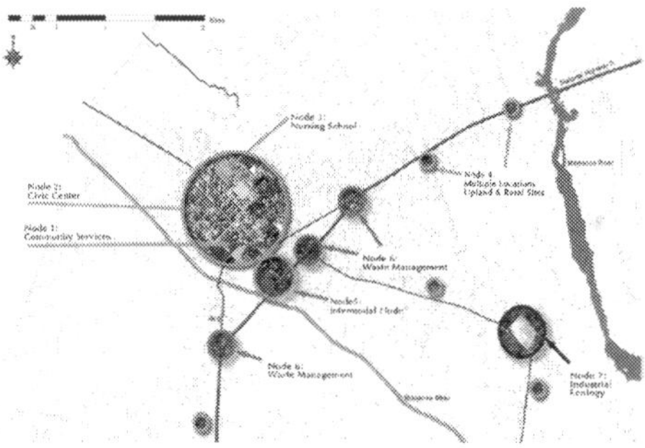

Figure 3: Downtown infrastructural nodes.

Proposal Components

Generation of Renewable Energy

Léogâne's renewable power system (with diesel back-up) will have its base load generated by PHS, powered by distributed wind and solar PV farms of 10 MW and 8.6 MW respectively. Electricity produced by these renewables will pump a quantity of water from the lower to the upper reservoir, where it is stored until released on demand (Figure 4). Passed through turbines, it generates the base-load for the microgrid. The PHS effectively acts as the storage battery and load-leveling device for the overall system (one of GEMi co-founder Daniel C. Gregory's many primary organizing concepts). Léogâne's PHS system is sized at a 10 MW, with a 180-meter head. The upper and lower reservoirs each have a water storage volume of 157,000 m^3. The closed-loop flow of PHS water through surface-mounted pipes will not interfere with local riverine ecology; however, these must be placed in stabilized areas. For 8–10 hours during the day, electricity will reliably feed local micro-grids serving key industrial, commercial and residential locations, as well as some outlying areas. This type of system, which minimizes transmission losses, could serve much of Haiti where mountainous regions and streams are adjacent to urbanized areas.

Figure 4: Energy system co-location benefits.

For system resiliency, the PHS system is reliant on multiple, complementary sources with intermittent energy supply. While Haiti's North and West Departments enjoy dependable wind velocities, Léogâne's flatlands do not reliably produce sufficient wind speeds. Placement of wind farms at higher inland elevations in the commune will better take advantage of seasonal and diurnal on- and off-shore breeze. One or more solar farms comprised of commercially available

PV panels will provide electricity to the PHS during the 11–13 hours of daylight. Excess power, not needed for water pumping, will augment the microgrid. Floating PV arrays are placed spanning across each of the PHS water reservoirs, comprising part of the generation system. Floating on pontoons, these arrays shade the water, reducing evaporation. The water in turn lowers the arrays' temperatures, improving their efficiency. The arrangement also avoids unnecessary displacement of agricultural land.

Placed on opposite hilltops at elevations of 600+ meters, a total of eighty-six 100 kW wind turbines will complement the solar farm as primary feed for the pumped-storage system. These medium size turbines, selected in part for easier transport on primitive roads, are supported by lattice towers that provide greater stability in this earthquake- and hurricane-prone region.

The GEM model assumes a long-term phased approach to 24/7 power. The electrical consumption planned per household is calculated initially as quite minimal; it covers basic energy services, allowing for night illumination, and the operation of small pumps for drinking water, radio and other small appliances including cell phone charging. Community nodes and hubs are more energy intense and require power sufficient for small refrigerators, laptop computers, basic medical equipment, and some income producing productive uses. In all, it is estimated that basic household services, along with commercial, community activities (schools and clinics) could be provided for on average just 50 kilowatt-hours per person per year, (Dilip Ahuja et al. 2008) for an annual project total of 5,000 MW hours.

Telecommunications Upgrades

Even before the 2010 earthquake, Haiti's telecommunications infrastructure was generally underdeveloped (fixed line penetration at the lowest in Latin America and the Caribbean). However, the country's internet connectivity is comparatively robust as most Haitian internet service providers connect to the internet via satellite and are not reliant on undersea fiber optic cables. Also, despite the national poverty level, mobile phone coverage has grown rapidly. For that reason, the community nodes will provide capacity for cell-phone recharging and servicing as well as internet cafes.

Ecological Restoration and Enhancement

That valuable natural resources lie within its boundaries is a condition hardly unique to Léogâne. Just as elsewhere across Haiti, these ecosystem services represent one of its greatest assets. These, unfortunately, have been subject to degradation. Therefore, environmental stabilization and resource regeneration are necessary preconditions to the development of energy systems, agricultural improvements, industry and economic development. Stormwater management, flood control, soil stabilization and reforestation are critical prerequisites to the installation of the energy corridor. A program of river stabilization (repair of eroding banks and other flood control measures constructed with local materials) will alleviate hazardous deluges that occur during tropical downpours. New reservoirs located upland will store stormwater for irrigation. Riparian buffering, described below, will help reinvigorate damaged marginal areas, while the introduction of new agriculture, agroforestry, aquaculture practices, and the new rural settlements supporting them, will stabilize the topography and restore hydrological systems.

The Delta

Restoration of the imperiled coral reefs and mangroves, along with flood mitigation and bank restoration, is imperative to re-provisioning ecosystem services. Mangrove restoration must take place first so as to protect fragile corals implants that act as filtering systems, trapping debris and silt, balancing nutrient loads. Mangroves protect coastal zones from the impacts of major weather events. Reforestation and riverbank reconstruction will further reduce outflows of sediments and must be advanced with the delta restoration. Secondly, coral reefs, which take decades to build, have a vital interconnected role and must be quickly cultivated to also serve as storm surge barriers, while they rebuild local biodiversity. A promising artificial platform system called the "Reef Ball"™ is recommended for reef regeneration (The Reef Ball Foundation-Designed Artificial Reefs). It consists of a perforated concrete form with plugs of healthy new coral fragment inserts placed on the ocean floor in suitable habitat which can be economically utilized to speed the repopulation process. Mangroves can be planted in biodegradable baskets using a range of salt-tolerant and saline resistant species based on local conditions.

Riparian Buffers

The riparian corridors along the Momence and Rouyonne were former oases of ecological diversity. Stripped of native vegetation, the eroded riverbank must be rebuilt to confine river flow. The middle zone, bordering the mountainous highlands and the alluvial plain, is ill-equipped to manage flooding, resulting in soil loss from agricultural lands and damage to human settlements. Riparian restoration within a demarcated corridor between 50 to 300 feet wide will recreate vegetated and forest buffer strips and patches that armor the riverbank to absorb floodwaters and slow erosion. Through pollutant removal and temperature moderation downstream, marine biodiversity and water quality will be restored. These new riverine habitat corridors will also aid in species migration across adjacent developed landscapes. During initial restoration, land holdings must be restructured along these rivers, including no-trespassing conservation zones and a less stringent long-term management system developed with local input.

Reforestation and Agroforestry

In addition to the soil retention and ecosystem services returned through restored riparian buffers, trees and plantings are reintegrated using 'agro-forestry' models. Proposed intercropping of trees with food plants will foster stewardship while increasing land productivity through crop-shading and nutrient exchanges. Fruit production, timber from jatropha trees, vertiver grasses and coir industries can be integrated with adjacent irrigation channels to boost agricultural productivity and local enterprise. The integration of agroforestry is both economically and ecologically vital for regeneration (Collier 2009). Local inhabitants must be able to benefit while enabling ecological services to flourish.

Upland Reservoirs

New reservoirs will be constructed at higher elevations in the energy corridor for flood mitigation and water storage. Independent of the pumped-hydro storage system, these reservoirs (875 m^3 and 220 m^3 capacity for the Momance and Rouyonne Rivers respectively), will divert floodwaters. Freshwater drawn from this impounded source will be also used for local crop irrigation. Micro-turbines placed within these channels will provide power to local farmers in rural areas not serviced by the micro-grid. New 100-foot riparian buffers within this upland zone, containing an estimated 380,000 trees, shrubs and grasses combined with riprap from concrete debris, will stabilize the soil.

Other Integrated Services

Transportation

As part of the integrated services development, improvements to roadbeds and associated storm water drainage are vital to support the installation and upkeep of new infrastructure and spur economic development. Concrete pavers such as ones currently used in Léogâne are recommended for surfacing main and side street roads and plazas in the city center, as these are pervious to water and easy to repair. For resurfacing peripheral roads, application of lime-based soil stabilizers renders a more reliable pavement with low environmental impact at low cost. Such improvements can serve heavier traffic, including small vehicles moving collected waste to a new 'eco-industrial park' (defined below).

One of the major service points also serves as a multi-modal transportation center: a dedicated hub located on one of Haiti's main highways just a short walk from the civic center. The station will accommodate buses, motorcycles, personal motor vehicles, and bicycles and has a small, attached market. It will be linked by new sidewalks, with pick-up and drop-off areas to improve pedestrian safety. Attached to the waiting area and public restrooms is a bio digester (see below) stationed to also receive large deposits of agro-waste delivered along the main road from outlying areas for transfer to the eco-industrial park.

Waste Management

Waste is an unrealized resource, its management central to local economic revitalization. It creates employment, fosters community involvement and promotes environmental stewardship. Proper refuse management reduces water supply contamination as well. A comprehensive proposal for Léogâne's waste diversion was developed that separates plastics, organics, metal and wood for recycling. It includes a new on-site processing facility, along with storage, transport, and connections to the large scaled biodigester and related generation facility (see An Eco-Industrial Park for Léogâne below). A mechanism to incentivize citizens to deliver sorted materials to the collection and pick-up points will be overseen by a new collection administration system. An educational campaign inculcating the benefits of resource conservation will also be vital for implementation.

Waste-to-biogas

Seventy-five percent of Haiti's total waste stream is organic, comprised of crop residuals and other biomass, along with food and other domestic waste (Booth et al. 2010). It constitutes a largely untapped resource. The plan calls for a distributed biogas production system made up of strategically placed and variously scaled bio digesters (small for a cluster of dwellings, to larger community-scaled units). These bio digesters rely on slow anaerobic digestion that yields methane from organic human and animal waste mixed with agro-residuals. This is converted into useful biogas and a residual soil amendment byproduct. Benefits include production of a new fuel source for cooking or for additional electricity generation; improved sanitation and public health; greenhouse gas reduction, and increased farm yields, along with job creation.

An Eco-industrial Park for Léogâne

Léogâne sits in an agricultural region long renowned for sugarcane production, Haiti's second largest cash crop after coffee. Yet in recent years, production and profits have faltered and Haiti has become a net importer of sugar from South America (Nienaber 2010). Léogâne now has one of the lowest yields for sugar cane in the Western hemisphere, with a range of 8 to 40 MT per hectare, compared to 85 MT in other parts of Haiti or Latin America (Barjon 2011b). Revival of the sugarcane industry (along with restoration of local food production) is essential for Haiti's overall recovery. Better management of existing resources and development of new markets would restore production to historic levels. A resurgent cane industry will restore much-needed jobs, as two-thirds of the Haitian work force and 25% of GDP production has remained in agriculture (Eberhard et al. 2010).

Léogâne's sugarcane production can be increased through more efficient management of existing resources as well as repair and expansion to the local processing facility, the Darbonne Sugar Mill. Additional resource efficiency can be gained through biodigestion of bagasse, the fibrous by-product of cane processing. The biogas extraction method has a higher energy capacity than simple bagasse combustion, which is a source of pollution. The process also co-produces a high quality fertilizer. Such an integrated resource management approach makes use of unrecognized assets, while reducing reliance on expensive fossil fuel fertilizers and potentially boosting agricultural yields (Piasecki 2013). Extensions to the Darbonne Mill property would effectively turn it into an eco-industrial park. Colocation of an industrial-scaled biodigester at the sugar mill would provide secondary or stand-by biogas cogeneration in Léogâne. The bagasse/biodigester upgrade can be coupled with Léogâne's collected organic waste, delivered and stored on-site. This new co-generation system would have a capacity of 31.1 MW (Brown & Ward 2012). Plastic waste would also be processed as a complementary use in this eco-industrial park, recycled into lightweight compressed bricks for local use.

DISCUSSION

Challenges and Next Steps

This proposal was built around the regulation of distributed renewable energy with the pumped-hydro storage as its lynchpin. To operate effectively, local environmental conditions must be stabilized. In addition construction must be staged to maximize resources. There are technological challenges as well. Pipelines must be protected from corrosion. The PHS water being exchanged requires filtration. The upland reservoirs will also need a measure of protection against sediment accumulation, a serious consequence of deforestation.

Successful ecological restoration will be contingent on temporary or permanent land acquisition. Stable administration and governance are essential for the necessary adjustments to land ownership and/or the creation of easements, complex social and financial transactions for Haiti. For fair distribution of services, placement of infrastructure assets cannot be done piecemeal, but rather systematically and according to a codified system such as GEM. Implementation of GEM's uniformity and modularity will require local, departmental and national cooperation.

A sophisticated understanding of local values and culture will be critical to program delivery. Particularly in Haiti, success or failure will depend upon the quality of local engagement as well as a commitment to local employment. The plan envisions an extensive stakeholder engagement process, which develops infrastructure based on input of local inhabitants, while conveying information about consumer responsibilities and costs. Long-term maintenance of these systems will demand skilled administrative and enforcement personnel. Therefore, a central strategy of GEMi, which will be necessary to implement this proposal, is to help train a workforce for the systems. Creating general awareness of the link between natural resources and infrastructure services is critical. This, along with technical training are GEMi cornerstones, not just for sustainable infrastructure, but ultimately sector-wide to balance short-term needs for water, food and shelter with investment in long-term. Given Haiti's history of electricity "piracy," it is also imperative to maintain the security of infrastructural components. The logistics of installing, maintaining and protecting expensive alternative energy infrastructure demand ongoing staff and security presence.

Creating and maintaining local political will and the capacity to coordinate with national ministries and other resources constitute the primary challenges to the plan's implementation. In-depth meetings with local leaders at all levels of governance and direct engagement with community members will be essential. Education will also be pivotal for acceptance of the paradigm shift involved in cross-sector integration. In many ways, Haitians are ready for this transition, given the failure of many infrastructure investments to date. A related challenge will also be first world buy-in to new infrastructural regimes implied by the integrated and autonomous eco district—one designed with context sensitivity, with local input and for local self-reliance.

Financing will be an obstacle for integrated infrastructure systems, as it requires coordination of resources and personnel from across what are traditionally 'siloed' ministries. Funds sourced from the international community are also needed for the initial outlays, while the program embeds plans for self-sufficiency after initial support is withdrawn. Estimations of lifecycle costs and comparisons between conventional fossil fuel (coal or gas) and the proposed alternative energy systems are not meaningful here since all fuel sources must be imported by this island nation. While currently, Petro Caribe provides a very generous rate to Haiti that allows for a 25 year payback period with 2% interest, (Jacome 2011), the Government of Haiti lacks sufficient credit to ensure a competitive fuel supply for the long-term. Costs for solar PV farms installed in Haiti would likely incur a 30 to 40% premium over U.S. costs due to charges for shipping, tariffs (Gregory 2011b) and transport to hard-to-access locations. Installation of a smart-grid energy network is more competitive for Haiti (as other emerging economies that lack existing energy infrastructure) and would provide a significant leap forward for this energy-impoverished nation. Biomass digesters are also likely to be cost effective given the tandem social and economic benefits of also providing a solid waste management system over and above a supplementary energy system. The cost benefits for natural systems regeneration are significant. The outlay for upland ecological restoration is likely to be more than offset by eliminating future loss of life and property from seasonal and hurricane flooding. The natural systems afford a measure of resilience where none currently exists.

CONCLUSIONS

Substandard infrastructure must be redressed in the recovery from major disasters and in the development of sustainable models (Kijewski-Correa et al. 2012). Numerous opportunities exist throughout the island of Haiti and elsewhere for replication of a "whole systems," integrated infrastructural approach. Most of the proposed technologies and strategies are adaptable to areas with available natural resources. In Haiti, where simply meeting the daily needs of the population is an extraordinary challenge, the creation of relatively low cost, integrated, self-reinforcing and resilient solutions can hopefully launch long-term self-reliance and economic development. Elsewhere, Small Island Developing States (SIDS) might utilize similar concepts, adapting them to local geography, meteorology and ecology as a pathway to sustainability. SIDS additionally face enormous stresses from climate instability. Therefore they can be at the forefront in adopting renewable technologies for both climate mitigation and adaptation purposes.

Vulnerability stems from inadequate knowledge as well as deficient resources to apply that information (Kijewski-Correa & Taflanidis 2011). This proposal builds capacity and resiliency through applied practical knowledge of resources, their interdependencies, and by optimizing their reciprocities. Various components of this proposal may find more universal application. Sector integration is both common sense and necessary in areas with limited resources for development. The restoration of ecosystem services, renewable power systems designed around the energy/water/waste nexus, and the use of modular and economic nodal service points, provide a powerful grounding strategy useful in many developing and even developed areas. Though yet untested, it represents a logical, and perhaps revolutionary next step towards sustainability from which the developed world too may draw lessons.

ACKNOWLEDGEMENTS

This project was conceived by the corresponding author and its components largely realized through her efforts and those of the teaching assistant working with Master of Science Urban Sustainability students participating in the seminar 'Case Studies in Sustainable Infrastructure' at the City College of New York. The design concept builds upon strategies under development by Daniel C. Gregory of Positive Energies, LLC who co-founded GEM.

REFERENCES

1. Adelman C (2011) Haiti: Testing the Limits of Government Aid and Philanthropy. Brown J World Aff 17(11):91
2. Archibold R (2010) Haitians Plunge into Muck to Stem Cholera. New York Times.
3. Barjon R (2011) Effectiveness of Aid in Haiti and How Private Investment Can Facilitate the Reconstruction. U.S Senate Subcommittees of Foreign Relations on International

Development and Foreign Assistance and Western Hemisphere hearing "Rebuilding Haiti in the Martelly Era". 15

4. Barjon R (2011) Effectiveness of Aid in Haiti and How Private Investment Can Facilitate the Reconstruction. U.S Senate Subcommittees of Foreign Relations on International Development and Foreign Assistance and Western Hemisphere hearing "Rebuilding Haiti in the Martelly Era.". 9
5. Barjon R (2011) Effectiveness of Aid in Haiti and How Private Investment Can Facilitate the Reconstruction. U.S Senate Subcommittees of Foreign Relations on International Development and Foreign Assistance and Western Hemisphere hearing "Rebuilding Haiti in the Martelly Era.". 6
6. Barjon R (2011) Effectiveness of Aid in Haiti and How Private Investment Can Facilitate the Reconstruction. U.S Senate Subcommittees of Foreign Relations on International Development and Foreign Assistance and Western Hemisphere hearing "Rebuilding Haiti in the Martelly Era.". 13
7. Booth S, Funk K, Haase S (2010) Haiti Waste-to-Energy Opportunity Analysis. NREL. 1 http://www.dinepa.gouv.ht/wash_cluster/index.php?option=com_rokdownloads&view=file&task=download&id=1025%3Anrel-haiti-waste-to-energy-final-nov-14-2010&Itemid=27
8. Brown H (2011) Global Energy Model's Pilot Project for Haiti: studies & concepts based on GEM's roadmap for energizing the country. 13
9. http://globalenergymodel.org/wp-content/uploads/2012/07/Haiti-Presentation-CUNY.pdf
10. Brown H, Ward M (2012) Planning an Ecodistrict: Integration of Critical Infrastructure Proposed for the Commune of Léogâne in Haiti. The City College of New York, CUNY. p 35
11. http://www.issuu.com/mnward/docs/gemi_leogane_booklet
12. Brown H, Ward M (2014) A Haitian 'ecodistrict:' conceptual 881 design for integrated, basic infrastructure for the commune of Léogâne. 882. Earth Perspectives 1:4
13. Cholera in Haiti. Center for Disease Control and Prevention. http://wwwnc.cdc.gov/travel/notices/watch/haiti-cholera
14. Collier P (2009) Haiti: From Natural Catastrophe to Economic

Security. Report for the Secretary-General of the United Nations. 17

15. Dilip Ahuja , Dilip , Tatsutani , Marika (2008) Sustainable Energy for Developing Countries. The Academy of Sciences for the Developing World. Sustainable Energy Rep. 21 http://twas.ictp.it/publications/twas-reports/SustainEnergyReport.pdf
16. Eberhard M, Baldridge S, Marshall J, Mooney W, Rix G (2010) The Mw 7.0 Haiti earthquake of January 12, 2010: USGS/EERI Advance Reconnaissance Team report. US Geological Surv Open-File Rep. 1048 iv
17. Global Energy Model Institute (2012) GEMi | Helping eliminate global energy poverty.http://globalenergymodel.org/
18. Gregory D (2010) The Global Energy Model. A Green Energy Corp White Paper. 10 http://greenenergycorp.com/File/View/b78fabc6-7247-483b-ac05-1253584d02d7
19. Gregory D (2011a) Global Energy Model. 4-6 http://globalenergymodel.org/wp-content/uploads/2012/07/GEM-WPv6.pdf
20. Gregory DC (2011b) Global Energy Model (GEM). Positive Energies, White Paper. http://globalenergymodel.org/wp-content/uploads/2012/07/GEM-WPv6.pdf
21. Haiti - Energy (2012) KOICA handed over two Electric Power Projects to the Government. Haiti Libre.http://www.haitilibre.com/en/news-6206-haiti-energy-koica-handed-over-two-electric-power-projects-to-the-government.html
22. Holm D (2005) Renewable Energy Future for the Developing World. In: International Solar Energy Society. Germany: ISES. p 2
23. Holm D, McIntosh J (2008) Renewable energy- the future for the developing world. Renew Energy Focus 9:1
24. Jacome F (2011) The Current Phase of Venezuela's Oil Dipolmacy in the Caribbean. Friedrich Ebert Stiftung. http://library.fes.de/pdf-files/bueros/la-seguridad/08723.pdf
25. Kijewski-Correa T, Taflanidis A (2011) The Haitian housing dilemma: can sustainability and hazard-resilience be achieved? Bull Earthquake Eng. 7 DOI: 10.1007/s10518-011-9330-y
26. Kijewski-Correa T, Taflanidis A, Mix D (2012) Empowerment Model for Sustainable Residential Reconstruction in Léogâne,

Haiti, after the January 2010 Earthquake. Leadersh Manag Eng 12:286

27. Klarreich K, Polman L (2012) The NGO Republic of Haiti. Nation. Koehler R, Mann P (2011) Field Observations from the January 12, 2010 Haiti Earthquake: Implications for Seismic Hazards and Future Post-Earthquake Reconnaissance Investigations in Alaska. 12 http://www.dggs.alaska.gov/webpubs/dggs/ri/text/ri2011_002.PDF
28. Lemons D (2012) At this writing, GEMi has been named by the Haitian Minister Delegate for Energy Security as the co-developer of a master plan for energizing the nation, fashioning a national grid based upon upgrades and additions to existing resources, implementation of new generation technologies, the roll-out of rural 'eco-districts', and the developer of a ten-year plan for ramping up education and training for the energy sector. Personal communication.
29. Library of Congress –Federal Research Division (2006) Country Profile: Haiti. 10 http://lcweb2.loc.gov/frd/cs/profiles/Haiti.pdf
30. Metzger J, Olsson A (2013) Sustainable Stockholm: Exploring Urban Sustainability in Europe's Greenest City. 200
31. Nienaber G (2010) Bill Clinton Puts Influential Muscle Behind Agricultural Production in Haiti. Huffington Post. http://www.huffingtonpost.com/georgianne-nienaber/bill-clinton-puts-influen_b_697906.html
32. O'Connor R (2012) Two Years Later, Haitian Earthquake Death Toll in Dispute. Columbia Rev.http://www.cjr.org/behind_the_news/one_year_later_haitian_earthqu.php?page=all
33. Piasecki M (2013) Personal communication.
34. OECD (2006a) Promoting Pro-poor Growth Infrastructure. http://www.oecd.org/development/povertyreduction/36301078.pdf
35. OECD (2006b) Promoting Pro-poor Growth Infrastructure. 13 http://www.oecd.org/development/povertyreduction/36301078.pdf
36. Rebuilding Léogâne Planning Workshop (2011a) Master Plan. "Partie 2: Plan de reconstruction post-séisme de la commune de Léogâne.". 2-8 http://www3.nd.edu/~ce4haiti/Information%20Database/MasterPlan.pdf

37. Rebuilding Léogâne Planning Workshop (2011b) Master Plan. "Partie 2: Plan de reconstruction post-séisme de la commune de Léogâne.". 2-2 http://www3.nd.edu/~ce4haiti/Information%20 Database/MasterPlan.pdf
38. Shepard W (2010) Streets of Haiti are Dangerous. http://www.vagabondjourney.com/streets-of-haiti-are-dangerous/
39. Sontag D (2012) Rebuilding in Haiti Lags After Billions in Post-Quake. New York Times.
40. The Reef Ball Foundation-Designed Artificial Reefs. http://www.reefball.org/
41. University of Notre Dame (2011) Rebuild Léogâne: Community Planning Workshop Report. 45 http://haiti.nd.edu/assets/54558/
42. University of Notre Dame (2011) Rebuild Léogâne: Community Planning Workshop Report. 27 http://haiti.nd.edu/assets/54558/
43. University of Notre Dame (2011c) Rebuild Léogâne: Community Planning Workshop Report. 18 http://haiti.nd.edu/assets/54558/

Chapter 8

Arsenic, Chromium, and Other Potentially Toxic Elements in the Rocks and Sediments of Oropos-Kalamos Basin, Attica, Greece

D. Alexakis[1] and D. Gamvroula[2]

[1]Laboratory of Reclamation Works and Water Resources Management, School of Rural and Surveying Engineering, National Technical University of Athens, 9 Iroon Polytechniou Street, 15773 Athens, Greece

[2]Institute of Mineralogy-Geology, Agricultural University of Athens, 75 Iera Odos, 11855 Athens, Greece

ABSTRACT

Rocks and sediments are non-anthropogenic sources of elements contamination. In this study, a series of potentially toxic elements were quantified in rocks and sediments of the Oropos-Kalamos basin. Only As, Hg, Pb, and Sb contents, in all the examined rocks and sediments, were higher than the levels given in international literature. Concentration of the elements As, Cr, Hg, Mo, Ni, and U is highly elevated in the lignite compared to crustal element averages. The enrichment of Cr and Ni in the lignite can be attributed to the known ultramafic rock masses surrounding the basin, while enrichment of As, Hg, Mo, Sb, and U is associated with the past geothermal activity of the Upper Miocene (about 15 million years ago). Nickel and Cr were transported into the lignite deposition basin by rivers and streams draining ultramafic rock bodies. The results of this study imply the natural source of Cr^{3+} and Cr^{6+} contamination of the Oropos-Kalamos groundwater, since high Cr contents were also recorded in the lignite (212.3 mg kg^{-1}), chromiferous iron ore occurrences (256.6 mg kg^{-1}), and alluvial deposits (212.5 mg kg^{-1}), indicating Cr leaching and transportation to the depositional basin dating from the Upper Miocene age.

INTRODUCTION

Many chemical compounds, major elements, and trace metals appear in waters and soils through agricultural, industrial, municipal activities, and natural processes [1–7]. In this way, there is an increase in the number of research studies dealing with degradation of water and soil quality [8–14], as well as with the levels of elements in vegetables [15] and biological samples [16]. Since, according to Salomon's0 [17], 99.9% of the metals are stored in soils and sediments, the high storage capacity of these materials plays an important role in controlling the fate and transpor0000000t of the metals. Based on these fundamentals, the investigation of the composition of soils-sediments-rocks is very important to determine the natural influences on water quality.

A previous study concerning the groundwater quality of the Oropos-Kalamos basin has been conducted by Stamatis et al. [14]. In that study elevated contents have been recorded for the following trace elements:

As, Co, Cr_{total}, Cr^{6+}, Mn, Pb, U, and V in the groundwater of Oropos-Kalamos basin. Furthermore, Stamatis et al. [14] applied speciation analysis and statistical treatment techniques in order to identify the factors controlling the elevated element contents in Oropos-Kalamos groundwater without, however, mentioning or investigating the composition of the representative solid materials (rocks, sediments, and soils) outcropping in the area studied.

The aim of this paper is, firstly, to evaluate the occurrence of trace elements in the representative samples from regional rocks and sediments, and, secondly, to define relationships between the trace element contents in the system: rock-sediment-groundwater.

Needless to say, that risk assessment of trace element contamination has become an important research topic dealing mainly with food quality and safety, health-related issues in humans, and environmental issues throughout the world. Recent developments in the field of risk assessment include research studies by:(i)Komnitsas and Modis [18], who assessed the risk related to As and Zn contamination in a coal mining region using variography analysis and numerical modelling;(ii) Komnitsas and Modis [19], who estimated the risk at waste disposal sites in the presence of hotspots;(iii)Vacca et al. [20], who investigated the total content of heavy metals in contaminated soils of the Rio Sitzerri floodplain (Sardinia, Italy) and discussed their impact on pedodiversity;(iv)Roy and McDonald [21], who studied the metal uptake in plants and assessed the health risk in metal-contaminated smelter soils;(v) Fernández-Calviño et al. [22], who studied the spatial variability of total copper, exchangeable copper, and bioavailable copper in a vineyard soil by applying statistical and geostatistical methods.

The results of this study will assist the local decision makers and regulators, firstly, to better understand the processes controlling the migration of trace elements into the Oropos-Kalamos basin and, secondly, to reduce environmental risk by selecting the best management practices.

MATERIALS AND METHODS

Description of the Study Area

The study area lies between latitudes of 38°14′–38°21′ and longitudes of 23°44′–23°52′ (Figure 1), being about 50 km north from Athens, the capital of the Greek Republic. It has an area of 80 km^2 approximately.

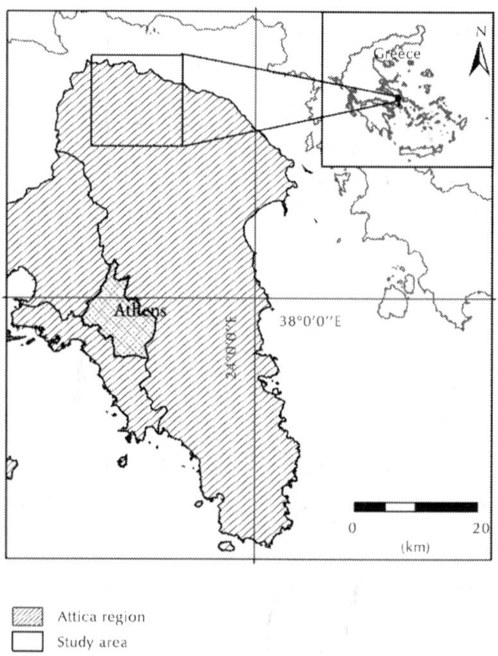

Figure 1: Location map of the area studied.

Geological formations which are observed in the study area include (Figure 2) schists, limestones, volcano-sedimentary formations, chromiferous iron ore occurrences, ultramafic rock masses, flysch, conglomerates, marly limestones with travertines, marly formations with lignite intercalations, and alluvial deposits [23–25]. The geology and lithology of the Oropos-Kalamos basin have been extensively presented in our previous study [14].

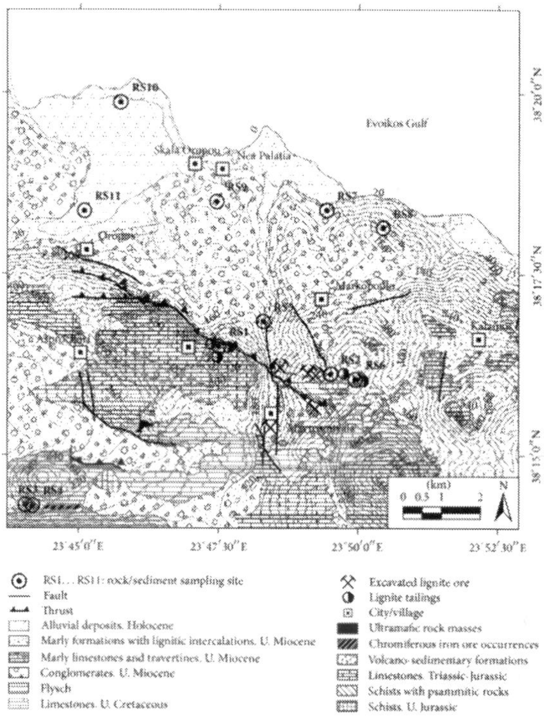

Figure 2: A simplified geological map of the Oropos-Kalamos area [23, 25] showing lignite tailings, excavated lignite ore occurrences, and sampling sites.

Corine land cover was used for obtaining the land use data [26]. The land use of the study area was classified into the following categories (Figure 3): (1) urban, (2) agricultural, (3) forest, shrub, and sparse vegetation, and (4) wetlands. The area studied includes the villages of Skala Oropou, Nea Palatia, Asprochori, Milesi, Mavrosouvala, and Kalamos and the cities of Oropos and Markopoulo. The agricultural land in Oropos-Kalamos basin is characterized by intensive agricultural activities especially vineyards, vegetables, and olive oil trees (Figure 3). Extensive lignite mining has occurred in the southeastern part of the study area since 1870 up to the 1960s. As a result of the mining activities, great amounts of waste produced from lignite mining activities were deposited in piles around the mining areas (Figure 3). There is a lack of industrial activities in Oropos-Kalamos basin.

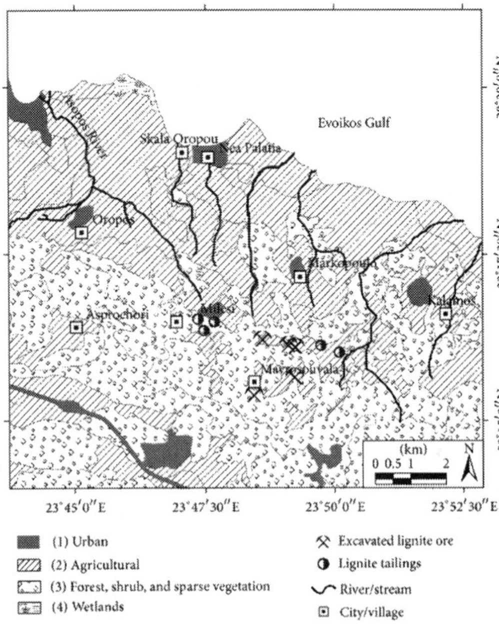

Figure 3: Map showing the land uses of the study area, rivers, streams, and names of the cities and villages.

Sampling and Laboratory Procedures

The rock-sediment sampling program was conducted with a total of eleven sampling sites (Figure 2). During the sampling process simultaneous site surveys were carried out in order to provide site specific information relating to industrial and agricultural activities near the sampling sites. Sampling sites were recorded in the field by using a Garmin eMap geographical positioning system (GPS). Rock and sediment bulk samples were collected from different lithological formations in order to be representative of the solid phases in contact with groundwater. Each sample collected comprised at least five subsamples that were homogenized into a single bulk sample which was about 2 kg. Specifically, two bulk samples were collected from lignites (RS1 and RS2), two bulk samples were taken from chromiferous iron ore occurrences (RS3 and RS4), two bulk samples were collected from marly formation (RS5 and RS6), three bulk samples were taken from conglomerates (RS7, RS8, and RS9), and two bulk samples were collected from alluvial deposits (RS10 and RS11) (Figure 2).

Bulk lignite samples were collected from the abandoned lignite mines vicinity. The surface of sampling sites was cleaned from vegetation and superficial debris before sampling by using a polypropylene tool and the samples were stored into clean polyethylene (PE) bags. Marly and conglomerate formation samples were taken from rock surface by using a hammer. The bulk lignite, chromiferous iron ore, marly, and conglomerate formation samples were air-dried at room temperature (<25°C), crushed, and grounded to <0.200 mm. Furthermore, the rock and sediment samples were quartered prior to chemical analysis. Contact with metals was avoided throughout samples preparation to avoid metal contamination. The homogenized and pulverized samples were digested with a mixed-acid attack employing concentrated HNO_3 and concentrated HCl. The resulting solutions were analyzed for Cd, Co, Cr, Cu, Fe, Mn, Pb, and Zn by atomic absorption spectroscopy method (AAS) using a GBC 908AA model at the Laboratory of Geology of the Agricultural University of Athens and for Ag, Al, As, Au, B, Ba, Bi, Ca, Ga, Hg, K, La, Mg, Mo, Na, Ni, P, S, Sb, Sc, Se, Sr, Te, Th, Ti, Tl, U, V, and W by inductively coupled plasma mass spectrometry (ICP-MS) at the ACME Analytical Laboratories Ltd., Vancouver, Canada (ISO 9002 Accredited Co.). Chemical analyses were monitored by using reference samples of the National Institute of Standards and Technology. Montana soil reference sample SRM 2710 was routinely analyzed to determine analytical errors (accuracy 5–9%). Triplicate measurements were conducted for 10% of the analyzed samples (precision 2–4%).

Homogenized and pulverized representative rock and sediment subsamples were made into pressed powder pellets and analyzed by powder X-ray diffraction (XRD) using an automated Siemens D5005 diffractometer. The scanning electron microscopic (SEM) analysis was used as a complementary technique with XRD analysis. Representative sediment and rock subsamples were examined by using a SEM model JEOL JSM-5600. An energy dispersive X-ray detector (EDX) Oxford Link ISIS 300 was used for the qualitative analysis of some spots of the rock and sediment samples. The XRD and SEM analysis were performed at the Laboratory of Economic Geology and Geochemistry, National and Kapodistrian University of Athens. Minerals were identified on the basis of morphology and SEM-EDX chemistry.

Statistical Treatment and Spatial Analysis

The geochemical data were treated using the Microsoft Excel software package. A 00000000000000) platform was created using ArcGIS 10.0 software (ESRI). The GIS platform includes basic layers showing the lithology, rivers, streams, and cities/villages locations. Sampling site locations, lignite tailings, and excavated lignite ore occurrences have been also plotted on additional information layers of the GIS.

RESULTS AND DISCUSSION

The mineral phases detected in the bulk lignite samples by XRD analyses are calcite, illite, gypsum, and montmorillonite. The XRD analysis of chromiferous iron ore occurrences was dominated by chlorite, illite, iron oxides, montmorillonite, and quartz, while ankerite and dolomite were detected in the marly formation samples.

The SEM observation and microanalysis results indicated the presence of calcite, organic matter, and pyrite in the bulk lignite samples as framboids and as isolated euhedral forms (Figure 4). The SEM examination of the chromiferous iron oxide samples shows the presence of chromite, hematite, illite, and quartz (Figure 5).

(a)

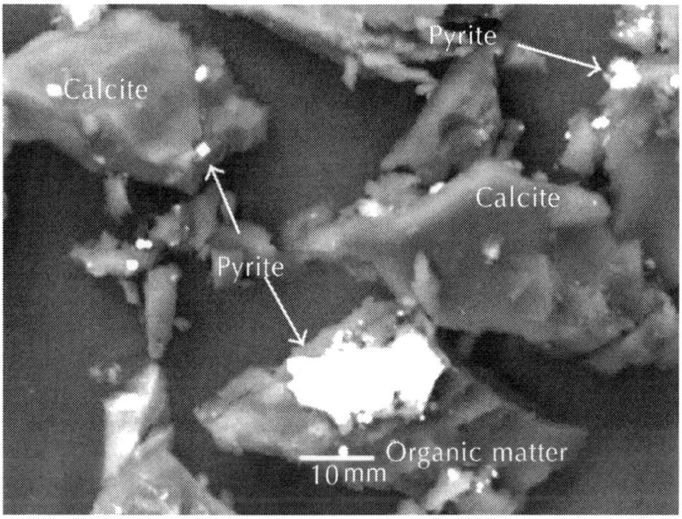

(b)

Figure 4: Modes of occurrence of pyrite in Oropos-Kalamos lignite of Upper Miocene age (SEM backscattered electron images). (a) Pyrite as framboid, organic matter, and quartz; (b) pyrite as euhedral, calcite, and organic matter.

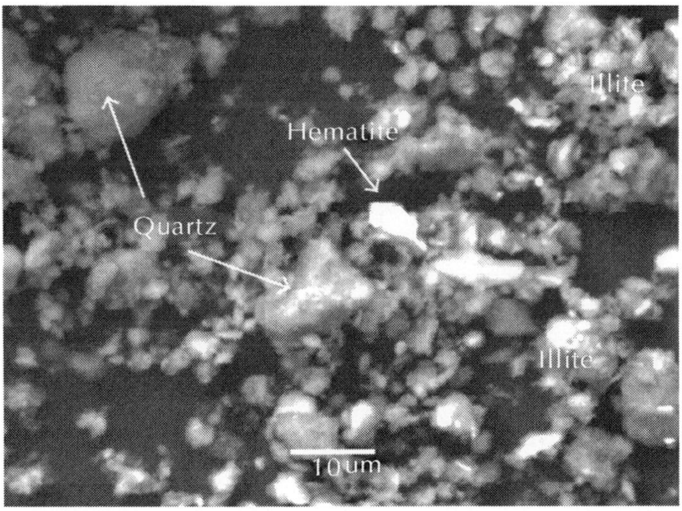

(a)

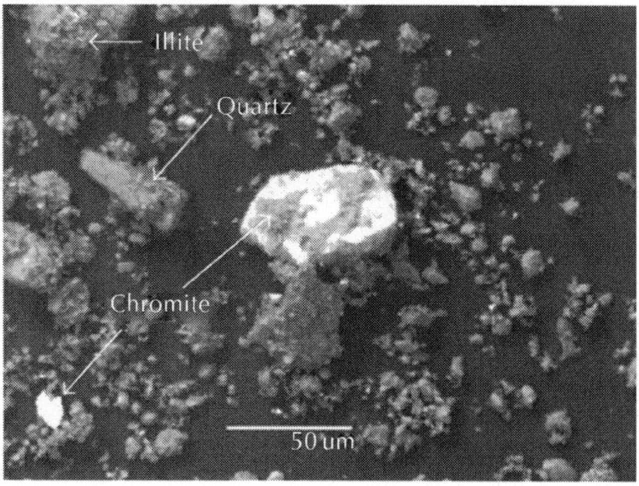

(b)

Figure 5: SEM images of chromiferous iron ore occurrences of Oropos-Kalamos area showing (a) hematite, illite, and quartz and (b) chromite, illite, and quartz.

Since Cr-Ni bearing minerals were not detected by XRD of bulk lignite samples, indicating that their concentrations are below the detection limits (±2 wt%) of the analytical technique, it is possible that some of the Cr and Ni have been transported into the lignite deposition basin as Cr^{3+} and Ni^{2+} in solution from ground or surface waters that drained the ultramafic rock masses at Oropos-Kalamos basin. Similar enrichment mechanisms have been postulated for Ni-rich lignite [27, 28]. Furthermore, high Cr and Ni contents of natural origin due to the weathering of ultramafic rock masses, ranging from 277 to 404 mg $kg^{-1}$and from 246 to 671 mg kg^{-1}, respectively, were also recorded at the coastal sediments of Evoikos Gulf [29].

The presence of pyrite as framboids and as isolated euhedral forms in Oropos-Kalamos bulk-lignite samples supports the hypothesis that pyrite is of syngenetic origin as a result of the reaction between ferrous ions, sulfur ions, and elemental sulfur by bacterial reduction in the reducing deposition environment. It could be suggested that the mechanism of pyrite (FeS_2) formation in the lignite of the study area is illustrated by the following reactions:

$$2Fe^{3+} + S^{2} \rightarrow S^{0} + 2Fe^{2+}, \qquad (1)$$

$$Fe^{2+} + HS^{-} \rightarrow FeS + H^{+}, \qquad (2)$$

$$FeS + S^{0} \rightarrow FeS_{2}. \qquad (3)$$

Furthermore, occurrences of pyrite in framboidal and euhedral forms have been also recorded in lignite deposits of Plakias-Lefkogia, Greece [30], and only as framboidal form in molassic sediments of the Evros area, Greece [31]. Pyrite of syngenetic origin in lignite is also reported for Yashan coalfield, China [32], and Megalopolis basin, Greece [33]. The absence of As from pyrite supports the hypothesis that As which is recorded in the bulk lignite samples of Oropos-Kalamos is mainly associated with the organic matter.

The mean values of the chemical composition of Oropos-Kalamos samples and the calculated enrichment factors are tabulated in Table 1.

Table 1: Mean values of the chemical composition of the examined samples in comparison with element contents obtained by other studies [34–36] and enrichment factors (EF)

	Units	DL	Bulk lignite	Chromiferous iron ore occurrences	Marly formation	Conglomerates	Alluvial	Elassona lignite mean [34]	World Coal [35]	Crustal average [36]	EF 1	EF 2	EF 3	EF 4	EF 5
Ag	μg kg^{-1}	2	44	87	10	14	—	—	—	—	—	—	—	—	—
Al	%	0.01	0.74	2.21	0.07	0.73	—	—	—	—	—	—	—	—	—
As	mg kg^{-1}	0.1	226.2	39.8	12.1	34.8	—	32	10	1.8	125.7	22.1	6.7	19.3	—
Au	μg kg^{-1}	0.2	2.9	2.1	0.9	1.9	—	—	—	—	—	—	—	—	—
B	mg kg^{-1}	20	56	20	24	20	—	—	—	—	—	—	—	—	—
Ba	mg kg^{-1}	0.5	30.7	135.8	58.2	60.5	—	—	200	—	—	—	—	—	—
Bi	mg kg^{-1}	0.02	0.2	0.33	0.02	0.09	—	—	—	—	—	—	—	—	—
Ca	%	0.01	2.01	2.35	20.69	13.86	—	—	—	—	—	—	—	—	—
Cd	mg kg^{-1}	0.01	0.69	0.55	0.02	0.1	0.01	—	—	—	—	—	—	—	—
Co	mg kg^{-1}	0.1	23.7	38.3	1.5	9.8	38.25	26	5	25	1.0	1.5	0.1	0.4	1.5
Cr	mg kg^{-1}	0.5	212.3	256.6	4.6	69.7	212.5	119	20	100	2.1	2.6	0.1	0.7	2.1
Cu	mg kg^{-1}	0.01	32.44	27.9	1.85	21.11	27.3	119	15	55	0.6	0.5	0.0	0.4	0.5
Fe	%	0.01	2.92	4.36	0.1	2.04	3.81	—	—	—	—	—	—	—	—
Ga	mg kg^{-1}	0.1	3.3	7.2	0.2	2	—	25	—	15	—	0.5	0.0	0.1	—
Hg	μg kg^{-1}	5	267	29.5	5	20	—	0.3	—	0.08	3337.5	368.8	62.5	250.0	—
K	%	0.01	0.06	0.3	0.02	0.07	—	—	—	—	—	—	—	—	—
La	mg kg^{-1}	0.5	9.9	22.3	0.5	5.8	—	77	—	30	0.3	0.7	0.0	0.2	—
Mg	%	0.01	0.33	0.51	9.99	0.71	—	—	—	—	—	—	—	—	—

Mn	mg kg⁻¹	1	94	1252	134	380	707.7	920	70	1000	0.1	1.3	0.1	0.4	0.7
Mo	mg kg⁻¹	0.01	7.63	0.51	0.12	0.24	—	15	3	1.5	5.1	0.3	0.1	0.2	—
Na	%	0.001	0.039	0.009	0.109	0.011	—	—	—	—	—	—	—	—	—
Ni	mg kg⁻¹	0.1	415.9	293.9	4.8	69.1	—	88	—	75	5.6	3.9	0.1	0.9	—
P	%	0.001	0.041	0.045	0.005	0.046	—	—	—	—	—	—	—	—	—
Pb	mg kg⁻¹	0.01	20.24	31.29	3.01	30.85	24.54	57	—	13	1.6	2.4	0.2	2.4	1.9
S	%	0.02	2.75	0.03	0.02	0.04	—	—	—	—	—	—	—	—	—
Sb	mg kg⁻¹	0.02	4.97	0.29	0.2	0.32	—	3	1	0.2	24.9	1.5	1.0	1.6	—
Sc	mg kg⁻¹	0.1	5.3	6.8	0.3	1.7	—	23	4	22	0.2	0.3	0.0	0.1	—
Se	mg kg⁻¹	0.1	4.3	0.8	0.3	0.6	—	—	1	—	—	—	—	—	—
Sr	mg kg⁻¹	0.5	365.4	13.4	910	123.6	—	608	—	375	1.0	0.0	2.4	0.3	—
Te	mg kg⁻¹	0.02	0.12	0.03	0.06	0.02	—	—	—	—	—	—	—	—	—
Th	mg kg⁻¹	0.1	1.8	5.6	0.2	1.4	—	25	4	7.2	0.3	0.8	0.0	0.2	—
Ti	%	0.001	0.011	0.014	0.001	0.008	—	—	—	—	—	—	—	—	—
Tl	mg kg⁻¹	0.02	1.61	0.3	0.03	0.05	—	—	—	—	—	—	—	—	—
U	mg kg⁻¹	0.1	8.8	0.5	1.4	0.3	—	7	2	1.8	4.9	0.3	0.8	0.2	—
V	mg kg⁻¹	2	225	65	3	14	—	187	40	135	1.7	0.5	0.0	0.1	—
W	mg kg⁻¹	0.1	4.2	0.1	0.1	0.1	—	—	—	—	—	—	—	—	—
Zn	mg kg⁻¹	0.1	38.9	72.4	4	63.52	92.8	176	50	70	0.6	1.0	0.1	0.9	1.3

The mean concentrations of As, Cr, Hg, Ni, Sb, U, and V in the Oropos-Kalamos lignite were significantly higher than those in the Elassona lignite given by Pentari et al. [34] (Table 1). Furthermore, the elements As, Co, Cr, Cu, Mn, Mo, Sb, Sc, Se, U, and V contents were higher than those in the World Coal reported by Clarke and Sloss [35]. Compared to the crustal average, the elements As, Cr, Hg, Mo, Ni, Sb, and U contents of lignite show enrichment factors varying between 2.1 and 3337.5 (Table 1). Enrichment factor (EF) is defined as the ratio of a measured element content to that of crustal average. The composition of the Earth's crust, rep.

orted by Mason and Moore [36], has been used for the calculations of enrichment factors. An element which shows enrichment factor higher than 2 can be considered enriched [34]. Enrichment factors higher than 2 are presented by using bold type characters (Table 1).

The enrichment of As, Hg, Mo, Sb, and U of the Oropos-Kalamos bulk-lignite samples can be attributed to geothermal activity that took place during the Upper Miocene in the Oropos-Kalamos basin, a fact that can be confirmed by the outcropping of travertine in the study area. Arsenic, Hg, Mo, Sb, and U enrichment by geothermal activity has been also reported for other Greek lignite, for example, Pentari et al. [34]. Moreover, the Cr and Ni enrichment of the bulk-lignite samples of Oropos-Kalamos basin is associated with ultramafic rock masses observed in the western part of the lignite deposition basin.

The elements As, Cr, Hg, Ni, and Pb concentrations appear to be highly enriched in the chromiferous iron ore occurrences, while the elements Co, Mn, Sb, and Zn present enrichment factors ranging from 1.0 to 1.5. The elements As, Hg, and Sr can be considered enriched in marly formation (Table 1). Furthermore, the elements As, Hg, and Pb contents in conglomerates can also be considered enriched, while the content of Sb is higher than that of the Earth's crust [36]. Chromium appears to be enriched in the alluvial deposits, while Co, Pb, and Zn show concentrations higher than the Earth's crust (enrichment factors between 1.3 and 1.9) (Table 1). Since there are no known industrial sources of trace element contamination in the study area, the anthropogenic origin of elevated element contents such as As, Cr, Cu, Hg, Mo, Pb, U, and V can be excluded.

In the study area, the heavy influence of the composition of the rocks and sediments on the groundwater quality is recorded by our previous study [14]. Furthermore, Stamatis et al. [14] reported elevated As (up to 246 μg kg^{-1}), Cr^{6+} (up to 28 μg kg^{-1}), Cr_{total} (up to 34 μg kg^{-1}), Fe (up to 2486 μg kg^{-1}), Mn (up to 132 μg kg^{-1}), U (up to 14 μg kg^{-1}), V (up to 19 μg kg^{-1}), and Zn (up to 974 μg kg^{-1}) concentrations in Oropos-Kalamos groundwater.

CONCLUSIONS

The syngenetic origin of pyrite in lignites of Oropos-Kalamos basin is supported by its presence as framboids and as isolated euhedral forms. The absence of As from pyrite supports the hypothesis that As is mainly associated with the organic matter of lignite. Concentrations of potentially hazardous elements such as As, Cr, Hg, Mo, Ni, Sb, and U are highly enriched in the Oropos-Kalamos lignite compared to crustal element averages. The mean contents of As, Cr, Sb, U, and V in the Oropos-Kalamos lignite were significantly higher than those reported for the Elassona lignite (Greece) and the World Coal. The distribution of trace elements in the Oropos-Kalamos lignite shows that the enrichment of Cr and Ni can be attributed to the known ultramafic rock masses surrounding the lignite basin, while enrichment of As, Hg, Mo, Sb, and U of Oropos-Kalamos lignite is associated with the past geothermal activity of Upper Miocene. Nickel and Cr were transported into the Oropos-Kalamos lignite deposition basin by rivers and streams draining ultramafic rock bodies. Therefore the high contents of As, U, and V in the Oropos-Kalamos lignite result in the high As, U, and V contents in the groundwater. The observation of elevated mean contents of As, Hg, Pb, and Sb in all the examined rock and sediments, in combination with the lack of known industrial source of contamination, indicates the natural source of contamination in the area studied.

The results of this study imply the natural source of Cr^{3+} and Cr^{6+} contamination of the Oropos-Kalamos groundwater since high chromium contents were recorded also in the Oropos-Kalamos lignite, chromiferous iron ore occurrences, and alluvial deposits, indicating Cr leaching and transportation by rivers and ephemeral streams to the depositional basin dating from the Upper Miocene age (about 15 million years ago).

Future work could focus on assessing the risk at the Oropos-Kalamos basin by applying a soil geochemical survey and delineating the spatial variability of the total element soil contents as well as their bioavailable fraction. The risks posed to humans, grazing animals, and other organisms from exposure to the concentrations of trace elements should be further investigated.

ACKNOWLEDGMENTS

The authors would like to thank the staff of the Department of Economic Geology and Geochemistry, National and Kapodistrian University of Athens, for cooperation during the XRD and SEM analyses.

REFERENCES

1. D. Alexei's, "Assessment of water quality in the Messolonghi-Etoliko and Neochorio region (West Greece) using hydrochemical and statistical analysis methods," Environmental Monitoring and Assessment, vol. 182, no. 1–4, pp. 397–413, 2011
2. D. Alexakis, D. Gotsis, and S. Giakoumakis, "Assessment of drainage water quality in pre- and post-irrigation seasons for supplemental irrigation use," Environmental Monitoring and Assessment, vol. 184, no. 8, pp. 5051–5063, 2012
3. V. Cappuyns and E. Slabbinck, "Occurrence of vanadium in Belgian and European alluvial soils,"Applied and Environmental Soil Science, vol. 2012, Article ID 979501, 12 pages, 2012.
4. D. Gamvroula, D. Alexakis, and G. Stamatis, "Diagnosis of groundwater quality and assessment of contamination sources in the Megara basin (Attica, Greece)," Arabian Journal of Geosciences, vol. 6, no. 7, pp. 2367–2381, 2013
5. K. Papadopoulou-Vrynioti, D. Alexakis, G. D. Bathrellos et al., "Distribution of trace elements in stream sediments of Arta plain (western Hellas): the influence of geomorphological parameters," Journal of Geochemical Exploration, vol. 134, pp. 17–26, 2013
6. V. A. Tsihrintzis, H. R. Fuentes, and R. K. Gadipudi, "GIS-aided modeling of nonpoint source pollution impacts on surface and ground waters," Water Resources Management, vol. 11, no. 3, pp. 207–218, 1997.

7. D. Alexakis, D. Gotsis, and S. Giakoumakis, "Evaluation of soil salinization in a Mediterranean site (Agoulinitsa district—West Greece)," Arabian Journal of Geosciences, 2014
8. D. Alexakis and G. Tsakiris, "Drought impacts on karstic spring annual water potential. Application on almyros (Crete) brackish spring," Desalination and Water Treatment, vol. 16, no. 1–3, pp. 229–237, 2010
9. G. C. G. dos Santos, G. S. Valladares, C. A. Abreu, O. A. de Camargo, and C. R. Grego, "Assessment of copper and zinc in soils of a Vineyard region in the state of São Paulo, Brazil," Applied and Environmental Soil Science, vol. 2013, Article ID 790795, 10 pages, 2013
10. A. Kelepertsis, D. Alexakis, and I. Kita, "Environmental geochemistry of soils and waters of Susaki area, Korinthos, Greece," Environmental Geochemistry and Health, vol. 23, no. 2, pp. 117–135, 2001
11. A. Loukas, "Surface water quantity and quality assessment in Pinios River, Thessaly, Greece,"Desalination, vol. 250, no. 1, pp. 266–273, 2010
12. A. H. Moghimi, J. Hamdan, J. Shamshuddin, A. W. Samsuri, and A. Abtahi, "Physicochemical properties and surface charge characteristics of arid soils in southeastern Iran," Applied and Environmental Soil Science, vol. 2013, Article ID 252861, 11 pages, 2013
13. K. Papadopoulou-Vrynioti, D. Alexakis, G. D. Bathrellos, H. D. Skilodimou, D. Vryniotis, and E. Vassiliades, "Environmental research and evaluation of agricultural soil of the Arta plain, western Hellas," Journal of Geochemical Exploration, vol. 136, pp. 84–92, 2014
14. G. Stamatis, D. Alexakis, D. Gamvroula, and G. Migiros, "Groundwater quality assessment in Oropos-Kalamos basin, Attica, Greece," Environmental Earth Sciences, vol. 64, no. 4, pp. 973–988, 2011
15. M. B. McBride, "Arsenic and lead uptake by vegetable crops grown on historically contaminated orchard soils," Applied and Environmental Soil Science, vol. 2013, Article ID 283472, 8 pages, 2013

16. D. Varrica, E. Tamburo, G. Dongarrà, and F. Sposito, "Trace elements in scalp hair of children chronically exposed to volcanic activity (Mt. Etna, Italy)," Science of the Total Environment, vol. 470-471, pp. 117–126, 2014
17. W. Salomons, "Biogeodynamics of contaminated sediments and soils: perspectives for future research,"Journal of Geochemical Exploration, vol. 62, no. 1–3, pp. 37–40, 1998
18. K. Komnitsas and K. Modis, "Soil risk assessment of As and Zn contamination in a coal mining region using geostatisretics," Science of the Total Environment, vol. 164, no. 2-3, pp. 1185–1190, 2009
19. K. Komnitsas and K. Modis, "Geostatistical risk estimation at waste disposal sites in the presence of hot spots," Journal of Hazardous Materials, vol. 371, no. 1–3, pp. 190–196, 2006
20. A. Vacca, M. R. Bianco, M. Murolo, and P. Violante, "Heavy metals in contaminated soils of the Rio Sitzerri floodplain (Sardinia, Italy): characterization and impact on pedodiversity," Land Degradation & Development, vol. 23, no. 4, pp. 350–364, 2012
21. M. Roy and L. McDonald, "Metal uptake in plants and health risk assessments in metal-contaminated smelter soils," Land Degradation & Development, 2013.
22. D. Fernández-Calviño, B. Garrido-Rodríguez, J. E. López-Periago, M. Paradelo, and M. Arias-Estévez, "Spatial distribution of copper fractions in a vineyard soil," Land Degradation & Development, vol. 24, no. 6, pp. 556–563, 2013
23. IGME, Eretria Sheet. Geological Map 1: 50000, Department of Geological Maps, Institute of Geology and Mineral Exploration, Athens, Greece, 2000.
24. G. Pe-Piper and D. Piper, the Igneous Rocks of Greece. The Anatomy of an Orogen, vol. 30 of Beitrage zur Regionalen Geologie der Erde, 2002.
25. G. Voreadis, "The tertiary lignite basin of Malakasa-Oropos," Geological and Geophysical Surveys, vol. 2, pp. 141–180, 1952 (Greek).

26. M. Bossard, J. Feranec, and J. Otahe, CORINE Land Cover Technical Guide: Addendum 2000, European Environment Agency, Copenhagen, Denmark, 2000.
27. A. E. Foscolos, F. Goodarzi, C. N. Koukouzas, and G. Hatziyannis, "Reconnaissance study of mineral matter and trace elements in Greek lignites," Chemical Geology, vol. 76, no. 1-2, pp. 107–130, 1989
28. L. Ruppert, R. Finkelman, E. Boti et al., "Origin and significance of high nickel and chromium concentrations in pliocene lignite of the Kosovo basin, Serbia," International Journal of Coal Geology, vol. 29, no. 4, pp. 235–258, 1996
29. M. O. Angelidis and M. Aloupi, "Geochemical study of coastal sediments influenced by river-transported pollution: southern Evoikos Gulf, Greece," Marine Pollution Bulletin, vol. 40, no. 1, pp. 77–82, 2000
30. N. Koukouzas and S. Skounakis, "Pyrite framboids in the lignite deposits of Plakia-Leufkogia (Rethymnon Crete)," Bulletin of the Geological Society of Greece, vol. 25, no. 2, pp. 193–201, 1991 (Greek).
31. K. Michailidis, G. Trontsios, and S. Sklavounos, "The occurrence of framboidal pyrite in the onshore molassic sediments of the Evros area, Thrace county, N.E. Greece," in Proceedings of the 2nd Hellenic-Bulgarian Symposium on the Geological and Physicogeographical Problems of the Rhodope Massif, Thessaloniki, Greece, 1980, (Greek).
32. S. Dai, D. Ren, Y. Zhou et al., "Mineralogy and geochemistry of a superhigh-organic-sulfur coal, Yanshan Coalfield, Yunnan, China: evidence for a volcanic ash component and influence by submarine exhalation," Chemical Geology, vol. 255, no. 1-2, pp. 182–194, 2008
33. G. Siavalas, M. Linou, A. Chatziapostolou, S. Kalaitzidis, H. Papaefthymiou, and K. Christanis, "Palaeoenvironment of seam I in the Marathousa lignite mine, Megalopolis basin (Southern Greece),"International Journal of Coal Geology, vol. 78, no. 4, pp. 233–248, 2009
34. D. Pentari, A. E. Foscolos, and V. Perdikatsis, "Trace element contents in the Domeniko lignite deposit, Elassona basin, Central

Greece," International Journal of Coal Geology, vol. 58, no. 4, pp. 261–268, 2004.

35. L. Clarke and L. Sloss, "Trace elements: emissions from coal combustion and gasification," International Energy Agency Report CR/49, London, UK, 1992.
36. B. Mason and C. B. Moore, Principles of Geochemistry, Wiley, New York, NY, USA, 1982.

Citations

CHAPTER 1

Claudia Santibañez, Luz María de la Fuente, Elena Bustamante, Sergio Silva, Pedro León-Lobos, and Rosanna Ginocchio, "Potential Use of Organic- and Hard-Rock Mine Wastes on Aided Phytostabilization of Large-Scale Mine Tailings under Semiarid Mediterranean Climatic Conditions: Short-Term Field Study," Applied and Environmental Soil Science, vol. 2012, Article ID 895817, 15 pages, 2012. doi:10.1155/2012/895817.

CHAPTER 2

Ruth Meinzen-Dick, Property Rights and Sustainable Irrigation: A Developing Country Perspective, doi:10.1016/j.agwat.2014.03.017.

CHAPTER 3

Nicolas Greggio, Pauline Mollema, Marco Antonellini and Giovanni Gabbianelli (2012). Irrigation Management in Coastal Zones to Prevent Soil and Groundwater Salinization, Resource Management for Sustainable Agriculture, Dr. Vikas Abrol (Ed.), ISBN: 978-953-51-0808-5, InTech, DOI: 10.5772/50534.

CHAPTER 4

Ouyang W, Chen S, Cai G, Hao F (2014) Dryland Soil Hydrological Processes and Their Impacts on the Nitrogen Balance in a Soil-Maize System of a Freeze-Thawing Agricultural Area. PLoS ONE 9(7): e101282. doi:10.1371/journal.pone.0101282

CHAPTER 5

McDonald RI, Girvetz EH (2013) Two Challenges for U.S. Irrigation Due to Climate Change: Increasing Irrigated Area in Wet States and Increasing Irrigation Rates in Dry States. PLoS ONE 8(6): e65589. doi:10.1371/journal.pone.0065589

CHAPTER 6

Laurel J. Lacher, Dale S. Turner, Bruce Gungle, Brooke M. Bushman and Holly E. Richter, Application of Hydrologic Tools and Monitoring to Support Managed Aquifer Recharge Decision Making in the Upper San Pedro River, Arizona, USA, doi:10.3390/w6113495

CHAPTER 7

Hillary Brown and Miriam N Ward, A Haitian 'ecodistrict:' Conceptual Design for Integrated, Basic Infrastructure for the Commune of Léogâne, doi:10.1186/2194-6434-1-4.

CHAPTER 8

D. Alexakis, and D. Gamvroula, Arsenic, Chromium, and Other Potentially Toxic Elements in the Rocks and Sediments of or opos-Kalamos Basin, Attica, Greece, http://dx.doi.org/10.1155/2014/718534.

Index

E

F

G

H

I

K

L

M

N

O

P